DOES GARLIC GROW ON TREES?

AN IRREVERENT GUIDE FOR BEGINNING GROWERS

BY

LEEANNE SANDERS

ILLUSTRATIONS BY CINDY BORDEN

ON THE LAM PRESS
LIVERMORE, COLORADO

Copyright © 2023 LeeAnne Sanders
Illustrations by Cindy Borden © Cindy Borden
Cover Design by Maura Kelpy, www.maurakcreative.com

All rights reserved. No part of this book may be reproduced in any form or by any means whatsoever, with the exception of brief passages embodied in critical reviews, without the expressed written consent of the publisher.

CONTENTS

PREFACE	1
CHAPTER 1: DOES GARLIC GROW ON TREES AND ARE VAMPIRES REAL?	5
CHAPTER 2: HARDNECK, SOFTNECK, GARLIC VARIETIES, AND OTHER THINGS WORTH GETTING IN A FISTFIGHT OVER	12
CHAPTER 3: BIG HEAD, PENETRATION DEPTH, AND DOING THE DIRTY	31
CHAPTER 4: WATER, WEED, FERTILIZE, REPEAT	42
CHAPTER 5: YOU DID IT!	46
CHAPTER 6: FUCK YOU, FUSARIUM	62
CHAPTER 7: USE THAT SHIT!	69
CONCLUSION: CONGRATS ON FINDING YOUR PLACE IN THE UNIVERSE	82
BIBLIOGRAPHY	86

Preface

There are better reference guides out there for growing garlic than this one. Certainly, there are more reverent and professional ones.

Neither reverence nor professionalism was a goal in writing this booklet.

Accessibility was.

Anyone can grow garlic. Anyone, too, can keep a garden if they're willing to do a little research and physical labor.

In a world where humans are becoming increasingly disconnected from the food they eat, making gardening approachable feels like a critical service. For new or hesitant gardeners, garlic is an excellent starting point as it's a relatively low-maintenance plant that's going to grow well 99% of the time.[1] Growing a great crop of garlic feels like a win. Gaining a gardening win leads to a little more confidence. A little more confidence leads to trying more things.

One year, you're growing a dozen heads of garlic. The next year, you're decked out in a flowered-print prairie dress, preaching the local food gospel while you sell kohlrabi from the back of a pickup truck on your neighborhood street, until the HOA cracks down on you because street vending isn't allowed and HOAs hate everyone

---

[1] Statistic completely invented but also, probably, somewhat accurate.

who digs up their yard and turns it into a garden. So you ditch the suburbs, move into a mobile home on a piece of property in the country, get a herd of milk goats, and start canning beets and making your own sauerkraut.

Garlic escalates quickly.

Or at least it did in our case.

A couple of suburbanites who grew up with zero agricultural experience and not so much as a backyard garden, my husband, Mike, and I both discovered gardening as adults. In fact, when we met on an online dating app over a decade ago, gardening was what we bonded over. Forget nudies and sexting. We got all hot and bothered over stories of Swiss chard and kale.

Yes, I did just reference sexting in a gardening booklet. This is a clear harbinger of things to come. If you are easily offended and do not find words like *asshat* hilarious, I recommend you bail now. I promised to make garlic growing accessible, not appropriate.

Online dirty talk about soil and compost quickly turned into an out-of-control gardening habit (and marriage), which subsequently escalated into a market farm growing a wide variety of produce—which, in turn, veered rather unintentionally into a small garlic farm.

As professional growers, however, our tenure was short-lived. Just a year and a half into market farming, Mike was diagnosed with multiple sclerosis, and after seven seasons of selling at our local farmers market, it became obvious that the workload had surpassed our decreasing ability to manage it.

So, naturally, we sold our suburban house, moved to a mobile home in the country, and continued to grow garlic (on a much smaller scale this time) and can our own beets and make our own sauerkraut.

We do not have a herd of milk goats, however.

Our neighbors do, so we just buy the milk from them.

No part of the above story is intended to garner sympathy or sell additional booklets to pay medical bills.[2] Instead, we want to motivate you into getting off your bunghole and actually growing something. If we managed a farm while also juggling a progressive, degenerative, neurological disease (and two off-farm day jobs), you are 100% capable of planting and harvesting a raised bed of garlic.

Which brings me back around to my original point that there are better reference guides out there. Many growers have been growing garlic for a lifetime. Others have done extensive research, some in conjunction with universities, to advance the understanding of garlic from a scientific point of view. Still others have written entire books on the subject of this stanky but popular plant.

We have done none of these things.

We began our garlic career by killing our first bed of garlic because Mike "didn't realize garlic needed to be watered," and we ended our career by becoming the second hit on Google if you searched for "Colorado Garlic Farm."[3]

We know what it's like to start growing from the ground up, and with the notes we took over the years, we have enough information to convey our learning without overwhelming the reader with deep dives into the genetic makeup of garlic, or detailed diagrams of the minutiae of its biology.

---

[2] Although this would be a really spectacular side benefit, so do, please, feel free to purchase numerous additional booklets.

[3] We're not anymore, so don't google it. It will just make me sad, and I'll miss it, and cry. But also, you won't find us because I took down our website so we wouldn't have to pay for it anymore because even if you are the second hit on Google for garlic, you still, most definitely, will not get rich. Also, those folks with the #1 hit must have been paying for hella SEO help. Never could beat them. But for posterity's sake in the annals of small-grower history, our farm was Lake Hollow Homestead.

We offer here simply a pragmatic take on growing garlic for new gardeners or aspiring market farmers, interspersed with a little fun and a lot of irreverence.

Just because you wear a prairie dress doesn't mean you have to be a prude.

CHAPTER 1

DOES GARLIC GROW ON TREES AND ARE VAMPIRES REAL?

Garlic is native to Central Asia, originating in a habitat with dry, harsh summers, brutal winters, and rocky, mountainous, inhospitable terrain.

In other words, if you fuck garlic up, you're a worse grower than a drought-stricken mountainside.

Encouragingly for novice growers, however, there's a difference between doing something (I can absolutely draw) and doing something well (I can absolutely draw stick figures only). Wild garlic growing in remote areas is spindly and small, looking nothing like the big, sexy bulbs found in backyard gardens and farmers markets.

Due to its scrappy origins, garlic is a surprisingly forgiving plant that will grow with minimal love and attention. Soil, water, and sunlight in reasonably sufficient quantities will net a decent harvest with little effort. Garlic is resistant to most pests, and while it is susceptible to a handful of diseases, it's not nearly as high maintenance as some plants that use any mild discomfort as an excuse to die.[4]

Unless you really are shittier than a remote Asian mountainside[5] (and if you are, no judgment here. You probably have other talents that we don't, like drawing non-stick figures), garlic is a solid entry-level crop that should grow, produce, and make you feel like a real-life gardener. We highly recommend it as an addition to any vegetable garden, especially those of new gardeners who are looking for an easy win. If you are one of those newbie gardeners, this pamphlet should provide some useful tips and tricks. Take the basics, ignore the rest, and you'll do just fine.

If you're an experienced gardener looking to up your garlic game, the nerdier details in this pamphlet are aimed at you. It's relatively easy to grow good garlic. It's a little more work to grow badass garlic. "A little more work," however, does not mean "not doable." It's totally doable.

Way more doable than drawing a human figure that doesn't look like it was created by a drunk preschooler. At least for us.

---

[4] I'm looking at you, tomatoes. Every summer, you're all "Ahh! Uneven watering…I have blossom end rot! Argh, it's too humid! Early blight! Now, it's too sunny! I sunburned! No wait…it's wet and cool! Late blight! You looked at me wrong! I'M DEAD."

[5] By "shitty" we are referring solely to the ideal conditions for growing a wide variety of vegetables. We are not implying inherent, overall shittiness of remote Asian mountainsides. In fact, remote Asian mountainsides are probably stunningly gorgeous. With beautiful stars at night. And few people. People are assholes. So, few people means few assholes, which means it's actually probably not shitty at all. Literally or figuratively.

So, let's get at it.

Does Garlic Grow on Trees?

Garlic is a vegetable.

Fun fact: fruits are foods that 1) develop from the flower of a plant and 2) hold seeds. If you're eating any other part of a plant, it's a vegetable.

If you would like to rapidly establish yourself as the local nutball, visit your local grocery store and quiz people on fruits and vegetables. Tomato? Fruit! Eggplant? Fruit! Rhubarb? Vegetable! If, go-getter that you are, you would like to up your game from local nutball to "Who is this crazy fuck, and would someone please get them out of my face before I call the cops?" start grabbing random fruits and vegetables from shelves and aggressively demand that people describe the foliage of the plants that produced them.

"I bet you can't tell me what a cantaloupe leaf looks like, can you, you useless, urban, I-live-in-a-condo-and-don't-cut-my-own-grass sonufabitch!"

While waiting to post bond, you'll have plenty of time to reflect on the fact that most Americans can't identify the foliage of even the common foods they eat, much less whether or not the food develops from a flowering structure on the plant or is part of the plant itself.

More than once at the farmers market stand, customers would approach our table and inquire, "So, how does garlic actually *grow*?" The most memorable gal, picking up a head of garlic and turning it around in her hand, said sheepishly, "I know this is a stupid question but... does garlic grow on trees?"

Not a stupid question. In a society where our food is sold in neatly organized displays and only the most picture-perfect specimens make it to the shelves, there's no reason to know what plants look like if you've never grown them yourself. While I'll happily and irreverently poke fun at almost anything, I'll never insult someone who is asking sincere questions to further their knowledge of gardening or agriculture. Or, frankly, asking sincere questions to further their knowledge of anything. Anyone who taunts anyone for trying to learn is an asshat.

But no, garlic doesn't grow on trees. It grows underground, producing long, narrow leaves above ground that spread out in something approximating a fan-like shape. Cultivated garlic is an allium (*Allium sativum*) and comes from the same genus that produces other pungent, edible plants such as onions, shallots, chives, and leeks.

Like potatoes, garlic plants are clones of the mother plant. When you split apart a head of garlic, each clove can be planted to produce a genetically identical plant. While some garlic varieties, mostly hardnecks, still possess the capability to produce a flowering structure called a scape, virtually no modern garlic is still capable of reproducing sexually. When a scape flowers, what appear to be seeds in the flower are actually called bulbils and, like the cloves, are clones of the mother plant.

While bulbils can be planted, it will usually take them two growing seasons or more to produce a full-sized head, so most garlic is propagated by planting cloves. The fact that cultivated garlic can no longer reproduce sexually indicates that over the thousands of years of cultivation, garlic has become dependent on humans for reproduction, banking on us to dig up those deliciously stinky bulbs, split them apart, and replant them with more space for each clove to grow.

Seriously, Though, WTF is the Deal with Garlic and Vampires?

The hippy-dippy world is rife with claims about how nature and its various plants can cure all ills. Some hippy shit is legitimate. Aloe does feel mighty fine on a burn, and lemon tea with honey works wonders on a sore throat. Lavender soothes, and ginger settles the stomach.

However, some of the hippy-dippy shit is simply snake oil sales masked in the modern-day guise of green smoothies, Birkenstocks, and patchouli for perfume. As the spouse of someone with multiple sclerosis, I know intimately the well-meaning but overbearing enthusiasm of natural remedy nuts. If one more person suggests that eating more kale and sweet potatoes might cure my husband's debilitating neurological disease, they're likely to get kicked in the nuts. Or the boobs. Depending on which part they have and if my husband's having a good-enough day to get his leg that high.

Hippy-dippy remedies should be backed up by actual scientific research. It is also true, however, that sometimes it takes scientific research a while to catch up with hippy-dippy remedies. This means it often feels like there's no real clear answer unless, by some miracle, the hippies and the scientists agree.

Garlic is this miracle.

A potent herb with substantial research behind its medicinal properties, garlic is both a culinary and medicinal rock star. Multiple scientific studies from reputable organizations indicate its effectiveness as an antimicrobial agent: it shows antibacterial activity, antifungal activity, and antiviral activity.[6] It shows effectiveness against

---

[6] Serge Ankri and David Mirelman, "Antimicrobial Properties of Allicin from Garlic," *Microbes and Infection* 1, no. 2 (1999): 125-129, https://doi.org/10.1016/S1286-4579(99)80003-3.

pneumonia,[7] streptococcus,[8] and even MRSA,[9] *Methicillin-resistant Staphylococcus aureus*, a super-bug that causes antibiotic-resistant staph infections.

Contrary to popular culture, however, garlic is not effective against vampires because vampires do not exist. That's it. They're a figment of the imagination. Of Brahm Stoker's imagination. Of Hollywood movies and teenage romance books.

Tracking down the origin of folklore is often difficult and almost always requires a liberal dose of conjecture, but the best connection between garlic and vampires generally comes down to the rabies theory.

Advanced by Gómez-Alonso in a 1998 study,[10] this theory connects the occurrence of widespread rabies outbreaks in the Balkans in the 1700s to "vampire outbreaks" that followed shortly thereafter. A number of the supposed symptoms of vampirism coincide with rabies: rabies sufferers become aggressive and often tend to bite, which causes contagion, as rabies is carried in the saliva. Spasms in the face and neck can cause teeth-baring and a snarling appearance.

---

[7] M. N. Palaksha, Mansoor Ahmed, and Sanjoy Das, "Antibacterial Activity of Garlic Extract on Streptomycin-Resistant *Staphylococcus Aureus* and *Escherichia Coli* Solely and in Synergism with Streptomycin,"
Journal of Natural Science, Biology, and Medicine 1, no.1 (2010), https://www.ncbi.nlm.nih.gov/pmc/articles/PMC3217283/.

[8] Kátia Andrea de Menezes Torres et al., "Garlic: An Alternative Treatment for Group B *Streptococcus*," *Microbiology Spectrum* 9, no. 3 (2021), https://doi.org/10.1128/Spectrum.00170-21.

[9] Guoliang Li et al., "Fresh Garlic Extract Enhances the Antimicrobial Activities of Antibiotics on Resistant Strains *in Vitro*,"
Jundishapur Journal of Microbiology 8, no. 5 (2015), https://doi.org/10.5812/jjm.14814.

[10] Juan Gómez-Alonso, "Rabies: A Possible Explanation for the Vampire Legend," *Neurology* 51, no. 3 (1998), https://doi.org/10.1212/WNL.51.3.856.

Rabies can also cause an aversion to strong smells, such as (you guessed it) garlic. Additionally, vampires are almost always portrayed as men, and rabies is more common in men than women. Vampires are also associated with animals that are or historically have been vectors of rabies outbreaks: wolves, dogs, and bats.

Since we can never go back in time to find the first dude (or dudette) who decided to sit around the family hearth and scare the shit out of the kids with stories of undead bloodsuckers suffering from an aversion to *Allium sativum*, we'll never really be certain about the garlic/vampire connection. Rabies seems like a solid theory, but in the end, it doesn't really matter because, once again, for those of you in the back of the room...

...vampires don't actually exist.

CHAPTER 2

HARDNECK, SOFTNECK, GARLIC VARIETIES, AND OTHER THINGS WORTH GETTING IN A FISTFIGHT OVER

Okey-dokey, folksies. The following fact is going to tick off some growers, even though they all know it because it comes from one of the best-publicized studies in the garlic world. No matter what anyone tells you, there are only ten types of garlic. That's it. That's all. In the whole universe—barring some as yet unknown, extraterrestrial garlic out beyond our galaxy.

Looking at seed magazines or grower sites, you would never know it, however. Garlic is categorized and sold under the names of hundreds of different varieties. Luckily, there are nerdy science people to keep us straight. In 2003, Dr. Gayle M. Volk of the National Center for Genetic Resources Preservation fingerprinted the DNA of 211 named varieties of garlic.[11]

---

[11] Gayle M. Volk and David Stern, "Phenotypic Characteristics of Ten Garlic Cultivars Grown at Different North American Locations," *HortScience* 44, no. 5 (2009), https://doi.org/10.21273/HORTSCI.44.5.1238.

HARDNECK, SOFTNECK, GARLIC VARIETIES, AND OTHER THINGS WORTH GETTING IN A FISTFIGHT OVER

Spoiler alert: ten. Did we mention that already? Dr. Volk found that there were just ten genetically different types of garlic. We'll get to those and their (fake? real?) sub-varieties in detail in a moment, but first, let's back up one more step.

All ten of the genetically distinct types of garlic fit into one of two categories: hardneck garlic or softneck garlic. While both hardneck and softneck garlic fit into the same species (*Allium sativum*, if we're speaking botanically), they show distinctly different characteristics.

Does all of this make your head spin? The taxonomy of garlic is a surprisingly contentious topic, fraught with dissenting opinions. As Ron Engeland puts it in his book *Growing Great Garlic*, "Professionals have wrestled with the problem of garlic varieties for over 100 years and basically succeeded in creating a very fine mess with very little agreement."[12]

Ted Jordan Meredith, in *The Complete Book of Garlic: A Guide for Gardeners, Growers, and Serious Cooks*, is slightly more diplomatic, stating, "The classification of garlic below the species level remains a challenging and problematic endeavor."[13] Meredith also includes an entire chapter on the taxonomy and diversity of garlic, and how the understanding of both has evolved. If you're a science nerd,[14] we suggest you read it. It'll make your head spin.

If you're just looking to grow some damn garlic, here's how we're gonna call it. This will be purely for the sake of being able to write

---

[12] Ron L. Engeland, *Growing Great Garlic: The Definitive Guide for Organic Gardeners and Small Farmers* (Okanogan, WA: Filaree Productions, 1994), 11.

[13] Ted Jordan Meredith, *The Complete Book of Garlic: A Guide for Gardeners, Growers, and Serious Cooks* (Portland, OR: Timber Press, 2008), 156.

[14] If you are unsure if you are a science nerd, read the following words out loud to yourself: *schema, germplasm, phenotypic plasticity, analytical model.* Are you titillated? Science nerd. Did you fall asleep? Not a science nerd.

a coherent booklet without losing our minds and getting mixed up in the barroom fight of a garlic taxonomy debate amongst nerds.

> **Species:** *Allium sativum*
>
> **Types of *Allium sativum*:** Hardneck and Softneck
>
> **Cultivars:** The ten types of garlic identified to be genetically distinct by Dr. Volk
>
> **Varieties:** The hundreds of names of garlic you see in seed pamphlets and at farmers markets

With that taxonomy question (not at all) settled, let's dig in.

Hardneck Garlic

Hardneck garlic is the darling of culinary snobs. Hardneck heads have a firm stem (literally, a hard neck) and fewer but larger cloves than softneck varieties.

The garlic you typically buy at the grocery store is softneck. It has large cloves on the outside, and then progressively smaller and smaller cloves running through the middle.

Now, imagine that all of those little, shitty, hard-to-peel cloves in the middle are gone, replaced by a narrow stem, and the outside cloves expand in size (but not in number) to take up their space.

That's hardneck garlic. Few cloves, but large cloves and super easy to peel.

Hardneck garlic also has a reputation for being stronger and spicier than softneck garlic, another characteristic that makes it a fan favorite in the culinary world. For real garlic aficionados, it's almost too good to be true. Huge cloves? Easy to peel? Great culinary taste? There must be a catch!

Of course there's a catch. Mother Nature isn't just going to hand you a perfect plant on a silver platter with no strings attached. This is the gal that invented survival of the fittest, the food chain, and spitting cobras. Consider, for a minute, that Mother Nature gets her kicks creating snakes that can hock a loogie of venom into your eyeballs.

There's no free ride on planet earth.

As awesome as hardneck garlic is, the catch is storage life.

Whereas most softneck garlics will store for a *minimum* of eight months, with the longest-storing varieties lasting up to a year or more, hardneck garlic is going to *max out* at eight months. Many varieties last much less time, some as few as just three months. Still, for those three to eight months, hardneck garlic is spectacular.

In addition to its flavor and clove characteristics, hardneck garlic still has a couple more tricks up its sleeve... er, stem... that makes it a lovely choice to grow.

First off, it's quite cold hardy. Actually, it's cold needy.

All garlic varieties require a process of *vernalization*: extended exposure to cold temperatures before a plant can grow to maturity.

Most hardneck garlic plants require a longer and colder vernalization period than softneck. They like the cold; they *need* the cold, and are amazingly hardy in the face of brutal winters. Hardneck varieties are the go-to varieties of gardeners in northern regions where softneck just can't cut it.

Finally, in its shameless bid to be the more popular garlic, hardneck garlic produces scapes in late spring. A scape is an edible flowering stem that essentially gives hardneck garlic two harvests: first the scapes, then a few weeks later the bulbs themselves.

A double harvest! Great cloves! Extreme cold hardiness! If it weren't for that pesky issue of storage life, hardneck would be the clear winner.

Or would it?

Enter the little shit of a sister, softneck garlic.

Softneck Garlic

Softneck garlic accounts for the majority of the garlic grown in the US.

But how? If hardneck garlic is so awesome and easy to use and has scapes and looks sexy and peels nice and...

The answer is industrial garlic production. Think about it. When was the last time you saw hardneck garlic in the grocery store? Probably never. Only recently have I seen it start to pop up very occasionally at specialty health-food-type stores. The rare times I do see it, it's usually out of season and the cloves are already getting soft. That's the problem with that pesky short(er) storage life.

Softneck garlic is that sneaky little sister who saw hardneck's one weak spot and capitalized on it. Big time. Garlic grown at the industrial scale is grown not for flavor but for characteristics that make it easy to produce and sell. These characteristics include resistance to disease, transportability, storage life, and other less-than-sexy reasons. Who fits the bill? Softneck. If you're looking to produce huge amounts of garlic to ship and sell all over the country, why in the world would you choose one with a short storage life and an obnoxious stem through the middle that you have to sort out?

However, to imply that softneck is just the homely sister chosen for her practical traits would be misleading. The softneck varieties grown at the industrial scale—typically California Early and California Late—are, quite frankly, nothing special. But the softneck varieties found at farmers markets, in seed catalogs, or gifted from your neighbor's thriving garden...these softnecks stand up to even the hardiest hardneck.

Softneck garlic can produce more garlic per square foot of growing space. Yes, it has little cloves running through the middle, but essentially, each head of garlic is solid garlic. There's no inedible stem that has to be removed. If you're growing garlic to dehydrate and produce garlic powder, softneck is an excellent choice. Take, for example, the garlic varieties from the Artichoke cultivar. These can grow to be absolutely massive in size, producing heads the size of the palm of your hand. All of softneck's other positive characteristics aside, it's pretty damn fun to pull a hand-sized head of garlic out of the garden.

While hardneck garlic gets points for being quite cold hardy, it's only fair to give softneck garlic the same props for being more heat tolerant. Heat tolerance in the garlic world is a relative thing, but as a rule, softneck garlic doesn't require as long or as cold a period of vernalization as hardneck garlic does.

This means that softneck varieties that can't quite stand up to Minnesota winters do quite well in the Mississippi heat. They're fine with the southern version of winters: cool, gray, and rainy is enough. Folks in the deep south and far north of the US (or even moving into Latin America and Canada) generally grow very different garlic varieties. In the in-between parts of the country, most of us can grow both. Even in Colorado, where we regularly get cold snaps well below zero in the winter, we've had many healthy crops of both hardneck and softneck varieties.

Finally, softneck garlic can be braided. The lack of a stem allows it to be braided together in strings of any length, and while this may seem like a rather artsy-fartsy selling point for a vegetable, it's actually just a literal selling point. Each year, when we made garlic braids to sell at our farmstand, we sold out. When we sold tickets to DIY garlic braiding workshops, we sold out. People love garlic braids. There is, apparently, something to be said for cute storage mechanisms for vegetables.

So, which garlic to plant?

Both. Obviously.

Back to That Damn Garlic Variety Problem

The question has been burning in your brain, keeping you up at night ever since you started reading this booklet, and making you doubt everything you thought you knew about garlic morals and honesty as a vegetable:

If Dr. Volk's study showed that there are only ten genetically different types of garlic, then how are there hundreds of different varieties in seed catalogs and at market stands?

The garlic world just ain't buyin' what Dr. Volk is selling. This is not to say that anyone thinks she's *wrong* per se. No one I know is taking up beef with her genetic analysis of garlic. Hell, the majority of us wouldn't even presume to genetically analyze our alliums. We'd rather just eat them. And because we eat them, many garlic growers are adamant that all of those hundreds of varieties do act differently and taste different from one another.

So, are there really likely only ten types of genetically different garlic? Yes.

Will garlic growers also swear that all of their (more than ten) different types of garlic are unique and wonderful beings that all have their own personalities and flavor profiles? Also, yes.

Are garlic genetics and taxonomy confusing enough for potentially both of these things to be true? Oh, hell yes. If even the nerdy science types can't agree on this stuff, you can't expect a bunch of grubby farmers and gardeners not to throw down over whether their Spanish Roja can kick your Bogatyr's ass.

Speaking from personal experience, we always grew three to four of the different cultivars identified by Dr. Volk, and six to eight different varieties of garlic within those cultivars. Because we wanted to know what we were talking about, we regularly did garlic tastings, using both raw garlic and cooked. Even within the same cultivars, the varieties did taste different, at least to us. But the differences are subtle. Garlic is garlic and it's powerful stuff, regardless of the variety you use.

When talking with people at our market stand, I always likened the taste difference to trying different tomato varieties. If you just grab any tomato out of the garden and bite into it, you'll say, "tastes like a tomato." However, if you slice up several different tomato varieties, line them up on a plate, and then sample them one right after another, you'll become aware of the differences in flavor and texture.

That's garlic. If you're cooking pasta and need to add a clove (or five) of garlic, just grab one, any one, and throw it in. It will taste like garlic.

If you want to be able to describe each garlic you grow in excruciating detail, then do a taste test. You'll smell so bad no one will stick around to listen to your overwhelmingly informed descriptions, but you will have satisfied your curiosity. Hanging out alone with your sulphuric garlic breath is a small price to pay for such wisdom.

Kickin' It with the Cultivars

Having completed your adventures in garlic tasting, you will now smell so bad that you can bank on at least a solid twenty-four hours of being friendless. Waiting out your self-inflicted garlic exile[15] is an excellent time to read up on the ten genetically distinct cultivars of garlic, because what else are you going to do? You smell like a gym-room sock left alone to its funk in a long-forgotten locker room corner.

---

[15] Fact: there are zero effective remedies for garlic stank-breath except time. If people on the interwebs tell you differently, they are full of lies. They're likely also the same people who think the earth is flat and that comb-overs actually trick us into thinking they're not bald.

Hardneck Cultivars

Of the ten garlic cultivars that exist, eight of them are hardneck, and the differences between several of these cultivars are notable. Don't just go buying your garlic all willy-nilly. That's how you end up with crappy harvests and rotten garlic, and then you show up at my market stand accusing me of selling bad seed when really you didn't read the instructions or do any research.

Do your research so I don't have to tell you to go suck it.

Asiatic

Asiatic garlic grows quickly and matures quickly, but stores poorly. It's like that kid in middle school who hits puberty before everyone else, rules the popular crowd, lives like hot shit through high school, then shows up to the ten-year reunion 100 lbs. overweight smelling like cigarette smoke, PBR, and pine-scented car air freshener.

Due to their fast-growing, hard-living lifestyle, Asiatic garlic is great garlic to plant for an early harvest that you want to eat right away.[16] In warm climates, Asiatics mature even earlier. The grower's rule of thumb is to harvest Asiatic varieties when the plants have just one to two brown leaves (more on leaf color in the **Harvesting** section). Leaving the garlic in the ground for too long shortens the storage life even further.

One note: Asiatic varieties do not actually smell like cigarettes, PBR, or pine-scented air fresheners. They smell like garlic. I shouldn't have to clarify this, but then someone out there would come at me all butthurt because they don't understand metaphors and their garlic smells like garlic and not a 7-Eleven.

---

[16] "Right away" meaning within four to six months—and six is pushing it.

Rocambole

Rocambole is the garlic diva. Generally considered to have the best flavor of all garlic, this is the garlic to try if you are a culinary nerd. Rocamboles generally produce eight to twelve cloves in a single layer around the stalk. They peel very easily, which reduces cursing during cooking.

However, like most divas, Rocamboles have their hang-ups. They get weird about overwatering and refuse to perform well in wet areas. They just sit there in the moisture, eyeballing you with disdain and whining about their feet getting wet. They require more cold for vernalization than other varieties, so they will likely also underperform in areas with warm winters.[17]

Like the Asiatics, they also don't store particularly well. They are, essentially, the Elton Johns of the garlic world: glamorous, delightful, pretty damn amazing... and also prone to spectacular temper tantrums if their plane is late or it rains too much.

Porcelain

Porcelains have a special place in our hearts. Way back in the day, when we had no idea what we were doing with garlic, Porcelains were the first variety we planted in our garden. For some reason, Mike thought garlic "didn't need to be watered" (we live in a semi-arid desert climate) and didn't put them on drip. Shockingly, they all died.

Well, technically, they came up, produced shitty little heads, then went dormant. We could have pulled the shitty heads, re-planted them in the fall, and probably had a normal crop the following year, but as I mentioned, we did not know what we were doing.

---

[17] I'm looking at you, Deep South. Fifty-degree weather does not count as cold.

Garlic is tough and will survive a lot, but if you want the big, pretty, farmers-market-quality heads, water is needed.

Once we figured out the whole watering thing, we fell in love with Porcelains and they became our primary hardneck crop for all of our commercial years. We will likely continue growing them forever. Some growers find their taste to be on the sulfurous side, with too much of a bite, but those growers are weenies. As a group, Porcelains have the highest allicin content, allicin being the compound responsible for many of garlic's health benefits. They grow few but huge cloves, are easy to peel, adapt well to most areas, and are consistently solid performers. If they're not as glamorous as the Rocamboles, at least they won't demand Louis Vuitton handbags and a perfect climate.

One substantial drawback to Porcelains: their huge cloves mean that many varieties will produce only four to six cloves per head. On some of the most monstrous ones we've grown, we'll sometimes find just three cloves. When you are holding back your own seed stock[18] (which you should do because garlic seed is fucking expensive), you often have to hold back a quarter of your whole Porcelain crop to plant for the next year. It's a significant investment.

Purple Stripe

Purple Stripes are the grandpappy of all ten of the garlic cultivars. Purple Stripes are genetically the most similar to wild garlic and spend most of their days standing over all the other garlic cultivars with a corncob pipe in hand, muttering about the finicky needs of

---

[18] A note on the term *seed stock*: no garlic is grown from actual seeds unless you're a garlic scientist in a library doing garlicky research things. For growers, seed stock refers to the best-of-the-best cloves and heads held back to plant during the following season.

"kids these days." Purple stripes are named for the PURPLE FUCKING STRIPES in their skin coloration, and if you didn't see that coming, you probably should just throw in the towel now.

They are hardy, medium-storage garlics that do best with a solid cold snap in the winter so they can talk about the good ol' days on the Asian steppe and how they used to grow uphill through the snow both ways to school. Much like many grandpappies, these garlics are a little tight-fisted... er, tight-cloved... making them somewhat hard to peel. But if you want a traditional, salt-of-the-earth garlic, Purple Stripe is it.

Marbled Purple Stripe

The Marbled Purple Stripe garlics (wanna take any guesses on their coloration?) are similar to the Purple Stripes in appearance but they usually produce fewer, larger cloves. Marbled Purple Stripe takes after its Grandpappy Purple Stripe in that it is a rock-solid garlic that will perform well in a large variety of scenarios but probably won't win any taste-test competitions like the Rocamboles. One distinct benefit of Marbled Purple Stripe for growers in warmer climates is that as a hardneck that thrives in colder climates, Marbled Purple Stripe can also do well in warmer regions.

Glazed Purple Stripe

Glazed Purple Stripe is—yes, you guessed it—also known for its purplish striping and yes, it also has a glazed look to it. The wrappers often look like someone rubbed the thinnest sheen of varnish over them, though the coloration and "glaze" of this cultivar are greatly influenced by where and how they are grown. Glazed Purple Stripes love the cold and hate being overly wet. They are a little more tender and need a little more care when harvesting, curing, and storing.

Creole

Creole is the prince of garlic cultivars. As in "Little Red Corvette" Prince. Not "stuffy London royalty" prince. In fact, to quote Prince, Creole garlic is one "Sexy Motherfucker." It has vibrantly red and purple skins that make a grower shout things like "Oh, you sexy beast!" when pulling it out of the ground. It likes hot climates where it can live life wearing skimpy clothing and showing a lot of garlic skins.

Like Prince, Creole garlic is generally small.[19] Its lack of size, however, can be balanced by its overall toughness, drought tolerance, and willingness to grow and perform rockin' renditions of Prince songs even in the Texas heat.

Turban

Turbans are partners in crime to the Asiatic's fast-growing, short-living lifestyle. Like the Asiatics, they should be harvested at one to two brown leaves, and like the Asiatics, they will lose even more of their short storage life if left in the ground too long. That would be a damn shame because their shelf life, however, is already the shortest among these ten cultivars, maxing out at three to five months. Turbans do typically produce larger bulbs than the Asiatics, and fit in the same category of garlic to grow for a good early-season harvest. *Grow fast, live hard, get eaten young* is their motto. It's a terrible motto, but a solid short-storage garlic.

Softneck Cultivars

Compared to the variety of the eight hardneck cultivars, it might seem a weak start to say that there exist just two genetically dis-

---

[19] Prince measured in at a whopping 5'2", which is why he always wore platform heels. To date, Creole garlic has not yet been seen in platform heels.

tinct softneck garlic cultivars. Yet, these two cultivars represent the vast majority of garlic grown and consumed in the United States. As a rule, softneck garlics are longer-storing than hardneck garlics, and are generally less persnickety than some of the hardneck varieties like the prima-donna Rocamboles or the tender Glazed Purple Stripes.

The real benefit of softneck varieties? The soft neck! You can braid them! What's better than a cute, artsy, homegrown garlic braid hanging in your kitchen?

Probably world peace. Enough money for a secure retirement. Some rich relative picking up the tab to buy you a new car. Or a new house. Humans learning to coexist without being dicks.

But short of those few things, a garlic braid is pretty great.

Artichokes

The Artichoke cultivar is not an artichoke. It's garlic.

It is not related to an artichoke; it does not taste like an artichoke; it does not grow on an artichoke plant, which, by the way, is quite wild-looking. If you've never seen an artichoke plant, google it. Just for funsies.

The Artichoke cultivar inconveniently shares a name with a plant it does not represent. The reason—which I've actually never read anywhere but seems blatantly obvious after growing them—is that if you peel off the outer layer of skin on an Artichoke variety, the exposed cloves are arranged in what does look like a small artichoke.

Artichoke garlic is undeniably awesome. So awesome that it is the most commonly grown commercial garlic. And while all of you farmers-market-foodie types are having a heart attack over why a garlic grower would refer to a commercially grown garlic cultivar as "awesome" when everyone knows that grocery store garlic is decidedly un-awesome, I say to you...

...try some Inchelium Red. Or some Kettle River Giant.

Most of the commercial garlic is one of two Artichoke subvarieties: California Early or California Late. Both are crap. Despite the results of Dr. Volk's study indicating that there are just ten types of garlic, Artichoke varieties are one of the biggest convincing factors for me that real and tangible differences in the varieties of one cultivar exist. I cannot look at the little shit of a California Early from the grocery store in one hand, and a bigger-than-my-palm Inchelium Red in the other, and believe that they are the same thing.

Artichoke varieties are resilient, adaptable, grow large cloves, and some of the heads are downright ginormous. It's a fantastic garlic to grow for making garlic powder because you get so much more garlic per head, which translates into more garlic per square foot of growing space. You can braid them, eat them, store them, or you can wave them in front of people's faces while shouting, "I am a garlic master!"

Silverskins

Silverskin varieties produce smaller heads but more numerous cloves than Artichokes. They also tend to be stronger in flavor and store longer. In fact, of all the garlic cultivars out there, Silverskins are by far the best storage garlic, lasting up to a year or even more under ideal conditions. That's a long freaking time. Imagine if that tomato you pulled off the vine lasted a year. It would be weird. But somehow with garlic, it's not weird. *Yeah, I'll just pull out this food that's been sitting on my shelf for a year and eat it.*

Silverskins are reasonably hardy, though they can't take quite as cold of winters as the Artichokes. That being said, we grew Silverskins for years in the Front Range of Colorado, where every winter we'd get at least one cold snap well below zero. The Silverskins did fine every year. Their smaller heads are sometimes easier to work with in braids, and for this reason alone we grew Nootka Rose, an excellent braiding Silverskin. It ensured that we would always have attractive braids to sell at the farmers market.

Storage Life

The storage life of garlic not only varies widely by cultivar but is also influenced heavily by how the garlic is handled. Harvesting too early or too late impacts storage life. How you cure garlic impacts storage life. *Where* you cure it, how you clean it, the storage bins you choose, ambient temperatures, ambient humidity levels, whether or not you sang it a little lullaby before you packed it in baskets, and whether or not it liked that lullaby or thought your singing was crap—all of these things affect storage life.

Therefore, when reviewing the list of average shelf lives below, understand that these timelines are quite variable. These numbers are generally accepted as typical, but don't be surprised if you find that your garlic lasts much shorter or even much longer than these timelines.

Asiatics: 5–6 months

Rocamboles: 5–6 months

Porcelains: 8-ish months

Purple Stripe: 6–8 months

Marbled Purple Stripe: 6–8 months

Glazed Purple Stripe: 6–8 months

Creole: 8 months

Turban: 3–5 months

Artichoke: 8–10 months

Silverskins: 1 year

Don't Plant Shit You Buy at the Grocery Store

When selecting your garlic seed stock, don't be a douche. You're reading an entire booklet on growing garlic, and if, upon turning the last page, the takeaway is to march over to your local grocery store, grab a few heads of whatever is on the shelf, and stick it in the garden, I'm ashamed and embarrassed. You should burn this booklet because you've learned absolutely nothing, you're a disgrace to gardeners everywhere, and I wish you bland and tasteless food for the rest of your life.

You could buy some trash garlic, certainly. It might grow. It might not grow. But the variety will be lousy (California Early or California Late) and you'll have no idea how old it is, where it came from, or if it's carrying disease. You won't be supporting small farmers, which is always a bonus of buying good seed garlic. Nor will you be able to easily identify at harvest whether your garlic sucks because you purchased trash, or whether your garlic sucks because you're a trashy grower.[20]

The three best places to purchase garlic are 1) online from reputable growers, 2) from seed catalogs, and 3) from reputable growers at your local farmers market. Which of these to choose largely depends on your desired outcome.

Seed garlic is expensive. It ranges roughly $16–$22 a pound and, compared to most garden seeds, is quite expensive to ship. A pack of lettuce seeds weighs nothing compared to even just one head of garlic.

However, if you're looking to grow significant quantities of garlic, or are considering growing it to sell, I would lean toward one of these first two options, even factoring in the shipping cost. Seed catalogs vet their growers. They often require testing for diseases.

---

[20] If you purchased garlic at the grocery store, it's probably both.

And generally, if something goes wrong with your order—such as the garlic arriving damaged or diseased—you can call and resolve the issue.

Many of these advantages are also true for pro growers who sell online. If a garlic farm is selling its own stock, they're not required to do testing, but many do anyway. Plus, it's always in the interest of any garlic farmer to keep their fields disease-free, as many garlic diseases are hard to eradicate once established.

Still, not all online farms grow excellent garlic. Do your research. Read reviews. See how long a farm has been in business. You should know these things. If you don't know how to do basic research, that's not the point of this booklet. That's for another booklet, to be written some other day. One called *Do Your Homework, Dipshit*.

Or would that be too rude?

Regardless, if you're looking just to grow a small amount of garlic for your garden, your local farmers market is a great place to look for seed. See some amazing-looking garlic at a stand? Buy it and plant it. See rough-looking garlic at a stand? Do not buy it and do not plant it. The *Do Your Homework, Dipshit* instructions apply to this too. Chat up the farmer at the stand. Ask them questions. Generally, most small farmers are good at growing most things, but are experts at growing only a few. We've grown everything from Thai chilis to tomatillos, but we know garlic best. Ideally, you want to buy from a farmer who knows garlic best.

One last note:

If you're purchasing online, be aware that most garlic-specific growers take orders for fall-planted garlic in *January*. That means that if you wait until September, you're probably going to have a hard time finding quality seed. Every fall, we were inundated with calls from both gardeners and nurseries looking for last-minute seed stock.

You know what these people didn't do? Their homework. Dipshits.

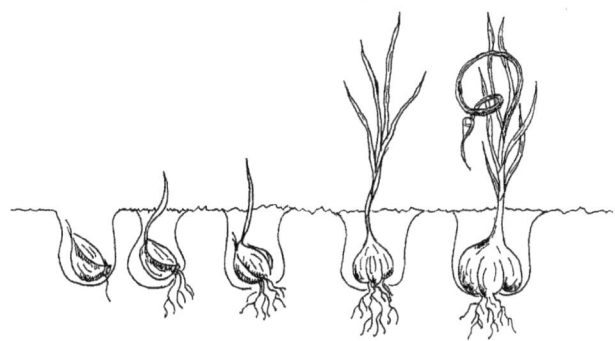

CHAPTER 3

BIG HEAD, PENETRATION DEPTH, AND DOING THE DIRTY

From here on, the primary purpose of this booklet is to share the garlic-growing tips and tricks that we learned in our years of growing for market. Much of the advice found here is universal and will translate well for any small grower in the US. However, like with all things gardening, another chunk of information will be directly influenced by growing region, soil, and other geographic variables that growers will need to learn to adjust for, either by seeking out resources specific to their regions or by... Doing Their Homework, Dipshits.

So, before launching into the details of our growing regimen, we'll outline some of the most notable variables of growing in our area: the semi-arid foothills of Colorado.

Rainfall

Colorado is bone dry.

It's not as dry as Arizona, say, or Nevada. But compared with much of the US, it is brutally dry. In our area, we receive fifteen inches of precipitation annually, and you can bet your bunghole that rain is not going to be evenly distributed. We get some rain in the spring, then summer is a total crapshoot, depending on if there are any afternoon thunderstorms. It's quite normal to go weeks, sometimes months on end with no rain. Without regular irrigation, the soil quickly becomes completely devoid of moisture. If you're growing in an area that receives more rainfall, you'll need to adjust the watering schedule (if you even water at all) as well as the monitoring of soil moisture.

Monitoring our soil moisture is easy: if we water, we can control it. If we don't water, everything dies. In areas with significant rainfall, controlling soil moisture can actually be more difficult. We can always add more moisture to our plots with irrigation, but to date, if it rains too much, there's no way to pick up your entire garden and wring it out.

Humidity

As with rainfall, you'll want to adjust for the moisture content in the air.

Mold, for us, is almost a nonexistent issue on the list of gardening struggles. We can hang our garlic in fairly large bunches without fans or a dehumidifier and still not have issues with mold. If you live in a humid climate, you'll want to plan the curing process to adjust for all that extra moisture content in the air. We'll get to that in the **Harvesting** section, but don't count on us for solid advice about drying garlic in wet climates, as that has not been our experience. No way, no how, no humidity.

Soil

Know your soil. This is basic advice for anyone growing anything, and it's another entry in the *Do Your Homework, Dipshit* booklet that, as of yet, does not actually exist. In our area of Colorado, we have clay soil. Clay has a lot of good stuff in it, as long as you can help the plants access it and loosen up some of its density with organic matter. The adjustments you make and the soil amendments you choose will vary based on your soil type, however. Sandy? Loamy? Rocky?

No idea what I'm talking about? Say it with me now: DYHDS.

Growing Zone

Where you live and what climate you have will affect what you can grow and how you can grow it. If you don't know it already, look up the USDA Hardiness Zone for your area. We're 5b. For garlic, this means we've been able to grow most cultivars of both hardneck and softneck just fine. For much of the US, this will be true. But you folks way north or way south may have a trickier time of it.

Planting

Garlic is a fall-planted crop.

Fall. Autumn. Ideally late autumn. *Fall* does not mean spring. Do not plant your garlic in the spring. To reach full maturity and produce the largest, healthiest heads, most garlic varieties need about nine months in the ground.

That's a long time. You can produce a miniature human in the time it takes a garlic clove to grow into a full-sized head. This blows many people's minds, as demonstrated by the eight-million, three-hundred-and-thirty-five conversations I had about it while working the farmers market. Condensing the aforementioned conversations down into one, it went something like this:

Customer: "I planted garlic this year. It didn't do crap."

Me: "When did you plant it?"

Customer: "Spring, duh."

Me: "Bummer. Should have planted it in the fall."

Customer: "Oh. That's an easy fix."

Me: "Yup."

In a perfect year, when we have our shit together, we aim to plant garlic by mid-October.

We do not have many perfect years and we virtually never have our shit together, so the garlic has gone in as early as mid-October (followed by fist-bumping, chest-thumping, and general proclamations of how awesome we are), and as late as late November (also followed by fist-bumping, chest-thumping, and general proclamations of how glad we are it's finally done). Every year, in all cases that the garlic has grown, we have gotten a crop. In the earlier plantings, however, the crop did notably better.

Remember that garlic requires an interval of exposure to cold temperatures. (We covered this in the **Hardneck Garlic** section with vernalization: extended exposure to cold temperatures.) When garlic is planted in late fall, the warm(ish) temperatures give it a chance to get a head start on the winter. The cloves throw down roots, which help stabilize them during the freezing and heaving of soil in the winter. They settle in, hunker down, and then, when the freezing temperatures hit, the garlic cloves go dormant. When things warm up again in the spring, garlic wakes up and resumes growing like winter never happened.

Anytime the ground isn't frozen, garlic roots are growing. Above around forty degrees Fahrenheit, the greens are growing as well. Garlic is a tough old bird... er, allium. But even hardy plants deserve a fair chance at life. So while you could, probably, just slap them in the ground in any old sunny location and be done with it, don't.

Better care makes for better garlic.

You need to get down and dirty and do things right.

Doing the Dirt(y)

In full transparency, any uptight gardener, farmer, or agricultural type will immediately correct you if you refer to your soil as *dirt*. "Dirt is what's on your floor before you mop it," they'll say. "Soil is what you grow nourishing food in." Then they'll take a sip of their high-end craft beer with a raised pinky and look at you judgingly over their glass with an equally raised eyebrow, assessing whether you are wearing the correct Carhartt pants and Blundstone boots for the task at hand.

Soil, dirt, ancient and pulverized rock powder—as long as you understand the basics of how it works, it doesn't matter what you call it. You can tell me you're planting your garlic in a pile of unicorn poo if that titillates your trowel.

Just make sure your dirt/soil/imaginary animal excrement is soft, fluffy, and well-balanced for your geographic region.

Defining what well-balanced soil means is a Pandora's box that I don't have the time or energy to tackle. This is not a booklet on creating healthy soil. If you're a nerd and you want to dig into soil pH, macro- and micro-nutrients, bioavailability, and the rest of the dirt on dirt, call your local extension office. Google it. Read another booklet that I have yet to write on soil amendments. DYHDS.

This section here is the quick and dirty. Or, rather, the soft and fluffy.

Garlic likes things comfortable. It needs to be planted at a depth of about four inches, and soft seed beds make that much easier. If you're gardening in raised garden beds, that may mean turning over the existing soil with a spade and then fluffing it up with a rake, or simply digging and loosening it with a trowel. If your beds have compacted over the years and the soil level has lowered, garlic planting is a great time to add some fresh soil or compost. If you're gardening in in-ground beds, tilling is by far the easiest solution.

As the science of gardening evolves, there has been a movement away from tilling as a tool for garden maintenance (or even as a tool at all.) No-till gardens certainly have real and important benefits, all of which originate from leaving the soil microbiology intact. In other words, when you don't disturb the soil, you make healthier soil, which then produces healthier plants and a healthier garden ecosystem overall.

However, when working with certain crops, namely underground ones like potatoes or garlic, disturbing the soil is just part of the deal. They're *underground*. So, in our years of growing garlic on a commercial scale, we rotated crops and worked hard at implementing no-till in the non-garlic beds, but continued to till the sections being planted in garlic. The light, fluffy base that tilling creates makes it easier to plant the garlic to the depth it needs to weather the winter and have room to grow and expand to full size.

Soft and fluffy soil grows sexy heads.

Poppin' Garlic

Before planting garlic, the heads must be broken apart into individual cloves, a process known as *popping garlic*. Every clove within a head will become an entirely new head, so if you just plunk a whole garlic in the ground without popping the cloves apart first, you're going to have overcrowded plants and small heads at harvest.

If you're just planting a handful of heads, you can just pop off the cloves as you plant. But if you're planting hundreds (or in our case, thousands) of heads, the prep time becomes significant and is worth doing ahead of time as its own chore. Garlic really should be popped as close to planting time as possible—the night before is perfect—but even doing it a few days ahead of time will likely be fine.

If popping a large quantity of garlic, and especially if you are planting multiple varieties of garlic that you want to ensure to keep separate, we recommend the following garlic-popping tips:

1. Do it outside. It's going to make a massive, papery mess.
2. Only pop one variety at a time. Otherwise, you risk mixing them up.
3. Pop each variety into a basket or paper bag and immediately label it.
4. Plan plenty of time.

Popping garlic is not as quick as one might assume. During our big planting years, when we were organized enough, we could make an event of it with trusted friends[21] and could knock out eight thousand-ish cloves in a few hours... with six to eight people working steadily during that time.

Of course, most backyard gardeners aren't going to be planting eight thousand heads because... well...

... that's a lot of head.

Selecting Cloves to Plant

Once your garlic is popped, consider whether you want to put all or just some of it into the ground.

---

[21] Trusted friends: those who would not get so schnockered on the adult beverages provided for the festivities that they might mix up garlic varieties, crush cloves, forget what they're doing, and/or spill beer on the garlic stash. If you do not pop garlic with alcohol involved, you may disregard this footnote.

As a general rule, larger cloves of garlic produce larger heads, so there is some argument for selecting the largest, nicest cloves of garlic as your seed stock. This is often done with softneck varieties, where growers may choose to plant only the largest outermost cloves and not the smaller inner cloves.

Hardneck varieties generally have fewer but larger cloves, so sorting out cloves by size is less common. Additionally, many hardneck varieties have so few cloves that growers generally want to plant everything they can from their hardneck stock; otherwise, it becomes quite expensive to hold back the following year's seed.

Over the years we grew, we always planted every single clove of our hardneck varieties. With the softneck we planted only the largest cloves, unless we had extra space. All cloves of a single garlic head carry the same genetics, so even the small cloves off a good-looking head will produce a decent-sized head the following year. If you have room and want to mess with breaking them up, stick 'em in the ground. If you're short on space, prioritize the big ones.

If you choose only to plant the largest softneck cloves for heads, don't throw the little ones away. These can be dehydrated to make garlic powder[22] (no, you don't have to peel them to do this), or you can plant them tightly together and harvest them as garlic greens in the spring. Technically, all parts of a garlic plant are edible, from leaves to roots. While the mature leaves of a garlic plant would be too tough and fibrous to be enjoyable eating, the young tender leaves can be harvested in spring and used in soups, pestos, or sautés for a light, garlicky flavor.

To plant a patch of garlic greens, prep some ground, roll those hunks of crappy little cloves between your palms to break up what you can, and scatter them over your planting space. You're packing

---

[22] Yes, I explain how to do this later in the booklet. I'm not such a big be-otch that I'd dangle dehydrated deliciousness in front of you without giving details.

them tightly and harvesting the greens as you would loose-leaf lettuce or another tightly grown crop, so perfection is not necessary.

If you end up with more garlic greens than you can use fresh, you can also dry these down and make a lighter-flavored garlic powder with them. Mixed with salt, this makes an excellent seasoning that isn't quite as potent as regular garlic salt, but adds a tasty depth to food.

Stickin' It In

With the soil prepped, garlic popped, and cloves selected, it's time to get down and dirty and stick it in.

Stick the garlic in, that is. It's dirty because you're kneeling in the dirt. It's time to get literally dirty. Not figuratively dirty. Though if you choose to get figuratively dirty while getting literally dirty, you certainly may. How you plant your garlic is entirely up to you.

But we recommend that you do it in rows, with plants spaced approximately six inches apart in all directions, planted at about four inches deep.

When planning out your garlic rows, it's easy to get excited and pack it into a bed. As you do, however, don't forget about weeding. In the spring, when the plants germinate and start to throw out leaves, you're going to have to weed between them. The lowest set of garlic leaves on a plant is usually quite near the ground, and that means that to weed, you're going to have to snake your hand in between them—however many rows deep they are.

Garlic plants can only produce a certain number of leaves a year. If they lose one, they can't replace it, and every leaf lost means a probable reduction in bulb size. Weeding through more than three rows of garlic spaced at six-inch intervals becomes difficult without damaging the plants, so limit your excitement and give yourself space to work.

Garlic should be planted at a depth of three to four inches. When planting, make sure you have the basal plate of the clove (the part that used to have roots attached to it) pointing down. Cover the planted cloves and water them well.

Mulching

Always mulch your soil.

Mulching means covering it, preferably with organic matter. Mulching the soil protects it from the elements, which destroy soil microbiology. When soil is exposed, the wind dries it out, then blows it away. The sun cooks it. The rain erodes it and creates gullies and runoff. Mulch insulates the garlic, protecting it from both temperature and moisture extremes, and thereby giving it a better chance of getting through the winter happy and healthy. Never leave your soil exposed, even if you're not growing garlic.

If you have garden beds you use in the summer, the worst thing you can do in the fall is clean them out then leave them bare. Leave the dead vegetable matter to protect them, then clean them out in spring, right before you plant.[23] You'll have much healthier soil, a better harvest, and, as an added bonus, the plant detritus creates an overwintering space for all sorts of bugs and insect larvae. Give those little critters a home. Right now, nature needs all the help she can get.

Over the years, our preferred choice for a mulching agent has been straw for a variety of reasons: 1) We have farmer friends who give it to us for free. 2) It's an organic material that, with snow, rain, sun, and time, will break down and create new soil. 3) We have clay

---

[23] One important caveat: if you know you have plants with contagious diseases that spread via plant detritus, those should be cleaned out and disposed of at the end of the growing season.

soil, and as the straw breaks down and mixes with the soil, it fluffs up the soil and makes the clay less dense.

However, any number of materials can be used: dead leaves, leftover hay, grass clippings.

Just keep that soil covered.

Drink Beer

Your work is done until spring. The garlic needs no fussing over, and any grower who gives you a detailed list of work to complete over the winter is an uptight ninny creating work where it doesn't need to exist.

Crack open a beer, pour yourself a whiskey, throw back a wine, and relax. Or do something that doesn't involve alcohol at all, though I don't really have any good suggestions for that. Sit there with a Shirley Temple and... um, sit there. Until spring.

CHAPTER 4

WATER, WEED, FERTILIZE, REPEAT

Sometime in March, you should lay off the beer (whiskey, wine) drinking and pull yourself together. Mostly because you don't want to become an alcoholic. But also because it's time to check on your garlic.

When the weather starts to warm, pull aside the mulch at intervals to look for tiny garlic shoots. If they're covered by mulch, they'll be yellow when they first pop up. As soon as you start seeing shoots, pull the mulch off all your garlic beds so the sunlight can hit them. Within a day or two, everything will quickly green up.

If you have a large garden, you might leave the mulch you've pulled off the garlic in the walkways between your rows. It will act as a weed suppressant and saves you the work of moving it elsewhere.

Then water, weed, fertilize, repeat.

Young garlic loves water. This took us a few years to realize because garlic is so damn hardy. It's damn near impossible to kill, but it also loves being pampered. Kind of like a really badass girlfriend. Tough and gritty when things are hard, but also kind of likes pedicures and long, soaking baths in fish fertilizer.

The garlic, not the girlfriend.

For the fish fertilizer, not the pedicure.

Garlic doesn't have feet. Duh.

If you live in a dry climate like ours, give garlic plenty of water from the time it sprouts (March-ish) until May or so.

Overwatering can cause disease and rot issues. However, this is much more likely in very wet areas. In Colorado, with our often brutally dry weather, it is hard for us to overwater. When we first started growing garlic, we set our drip irrigation at timed intervals like we always had done for vegetables. Depending on the weather, we might have had it on for ten minutes twice a day, or thirty minutes every other day, but always using some mathematical formula.

Then, in his incessant research around garlic, Mike read somewhere that garlic was like any other green in the spring and loved water.[24] We switched our spring watering from timed intervals to watering enough to keep the soil consistently moist (but not soaked), and our garlic responded instantly. It went from looking like the cute girl next door (which is not a bad look at all) to looking like a sleek supermodel strutting the garlic rows. It grew taller, stood straighter, and tossed its leaves around in the breeze like it was the sexiest plant in the garden. That garlic *owned* the garden that year.

---

[24] I would love to cite here the source of that interesting tidbit of information, but Mike has no idea where he read it. As Mike is generally researching something during 98% of his waking hours, factoids about garlic are likely to be interspersed with fascinating but irrelevant tips on building wooden boats in one's garage, techniques for maintaining perfect form on a rowing machine, or tidbits about the origin of the Cartman character in South Park.

We had never seen such a confident crop of garlic.

Every year after, we continued the increased water regimen and consistently reaped the same positive results.

Fertilization

At least once per month, during the same window of March to April, Mike would give the plants a heavy watering with fish fertilizer.

If you are unfamiliar with fish fertilizer, it is exactly what it sounds like: a fertilizer made from fish, and usually from the less desirable parts of fish such as the scales, bones, and skin. These are pulverized and then mixed into a sort of liquified fish goo. Like any synthetic fertilizer you buy at a garden shop, this fish goo can then be added to water to give plants a dose of nutrients when needed.

Fish fertilizer smells exactly as one would anticipate: like a restaurant's worth of leftover seafood left out in the sun to rot. Post-application, it gives the garden a nice fish funk for a solid week, an olfactory delight that your neighbors will appreciate—though perhaps not as much as your dog, who will happily roll away the hours in fish-scented mud and then come barreling in the house, smelling for all the world like they just popped out of a whale's ass before settling in on your nice clean couch for an odorous nap.

Apply accordingly.

Fish fertilizer is by no means the only fertilization method for garlic, but after years of experimenting—always with organic, non-synthetic options—it is the one we found worked best for our crops in our climate. Some garlic growers compost heavily at fall planting, then do nothing in the spring. Others swear by foliar feeding, a process in which a water/fertilizer mix is sprayed directly onto the plants rather than the soil. Regardless of the fertilization method you choose, any fertilization schedule should wrap up around early

May. Continuing to fertilize after that may result in decreased bulb size.

Experiment, play with it, test your own theories, but a deep watering with fish fertilizer, once in March and again in April, worked wonders for our crops. Even if, occasionally, the dog found the bottle and had herself a little fishy snack, resulting in multiple days of piscine dog farts.

A good garlic harvest is worth fishy-smelling dog ass.

Weeding

Garlic is not a competitive crop.

If you were playing Monopoly with garlic, it would just hand you Broadway and Park Avenue and then go live in a tent somewhere unobtrusive, where it could be left alone and sip on its fish fertilizer in peace. It has no interest in outwitting you to put some crummy hotel on a silly piece of land, nor does it feel like putting in the time or effort required to outcompete those overachieving weeds for an extra dose of water or a fraction of more sunlight.

If crowded out by weeds, garlic plants will be less vigorous and produce smaller heads, especially if competing with particularly aggressive ground-cover weeds like bindweed. Additionally, due to the long, narrow nature of their leaves, garlic plants don't compete well with weeds that might shade them. There's simply not enough surface area to make up for lost sunlight.

Give your garlic some breathing room and work off all that winter beer you drank.

Pull some weeds.

CHAPTER 5

YOU DID IT!

Scape Season

If you are growing strongly bolting hardneck varieties (typically Purple Stripe, Glazed Purple Stripe, Rocamboles, and Porcelains[25]), you'll have a harvest before your harvest: the garlic scapes.

In terms of arguments on this planet that don't actually matter, the garlic growers' argument over scapes ranks just a slight step above which way the toilet paper goes on a dispenser. Most growers maintain that cutting off the scapes on most varieties produces a larger head at harvests. A vocal minority claims that with proper care, it really doesn't matter all that much.

---

[25] Sometimes the other hardneck varieties may also produce scapes, though not as reliably, and occasionally softneck will throw out a scape or two, albeit mostly when the plants are stressed.

Our opinion:

Scapes are delicious and harvesting them might get you a better garlic crop, so why wouldn't we harvest them?[26]

Scapes will come in about a month before the heads are fully mature. In our growing region, that means the first week of June-ish.

If you've never seen a garlic scape, you'll recognize it right away. Scapes look quite different from the garlic leaves, appearing as firm, stem-looking things with a pointy flower bulb on the end.

Once they emerge, keep an eye on them. Scapes are sketchy sonsabitches. Left unsupervised, they might pillage your village, kidnap all the women and children, and burn your garden to a charred crisp, salting the earth behind them.

Or, in less drastic scenarios, they'll just grow too quickly and ruin your chances for a good garlic crop.

Same-same.

The trick with scapes is to cut them at the right time. Cut them off too soon, and they'll just grow back, wasting the plant's energy in repeating the same task twice. Cut them off too late, and they'll already have stunted the growth of the main garlic bulb.

When scapes grow out of the main plant, they will initially grow in a tight circle. If not cut, this curlicue will eventually straighten out and the stalk will shoot straight up into the air, reaching heights of six to eight feet.

You don't want this to happen. At this point, they've pillaged your village and decimated your garlic heads.

---

[26] Our enthusiasm for scapes aside, we are firmly on the harvest-for-a-better-crop team. Despite our best attempts at harvesting every single scape, every year we miss a handful and every year, without fail, the plants with scapes still attached have much smaller heads. That being said, in terms of hardneck, we grow primarily (though not solely) Porcelains.

The rule of thumb with scapes is to wait until they curl around once, then cut them at the base where they connect with the stalk of the plant.

Fresh scapes have a distinct garlic taste without the bite garlic cloves have. They are excellent for cooking. Use them as you would garlic in recipes. They can be ground into a garlicky pesto or chopped into a stir fry. They are great for seasoning soups or mincing into a marinade.

If you find yourself drowning in too many scapes, they can also be frozen, although this will change their consistency when thawed. Frozen scapes are best used as a minced or blended garlic seasoning. For the more ambitious, scapes can be dehydrated and ground to produce a milder-tasting garlic salt than that made with garlic cloves. Check out our Spring Salt recipe at the end of this booklet for our favorite scape use.

With the garlic grown, scapes harvested, and the first official days of summer on the horizon, it's game on.

Time to harvest some heads.

Harvest

We harvest garlic when it has around six green leaves left. For our region, this is usually somewhere between the last week in June and the first two weeks of July.

If you irrigate your garlic as we do, you'll want to cut off your irrigation about two weeks before your anticipated harvest to allow the garlic to start curing (drying out and creating those protective garlic wrappers) in the ground. If you live in a wetter region than we do, then, in addition to counting garlic leaves as they dry, you'll want to watch the weather for a dry streak—or as dry a streak as you can get—before harvesting your garlic.

We cut off water once scape harvest is finished. Depending on the heat, moisture, and general welfare of the garlic, we often end up pulling our hardnecks at the end of June, followed by our softnecks during the first week in July, and always ending with the Silverskins around the second week of July.

Harvest Prep

If you're a home gardener growing a handful of heads, you can ignore this section. Or you can read it for the sheer delight of making this spectacular booklet on garlic last longer. Or you can totally over-plan your small harvest and make it way harder than needed by prepping, labeling, and cataloging your ten heads of garlic from the garden, cementing your reputation as an uptight over-doer.

Your call.

If you're growing hundreds (or thousands) of heads and multiple varieties of garlic, harvest warrants some planning. Remember that outside of the hardneck/softneck question, most garlic varieties aren't easily distinguishable by sight. If you're well-versed in garlic, you may be able to tell the difference between a Porcelain and a Silverskin at a glance, but distinguishing between a Purple Stripe and a Marbled Purple Stripe will be nearly impossible. Determining the different varieties within a single family is, actually, impossible.

Since garlic plants are clones, if you mix up one variety one year, you're completely fuckered for the rest of your garlic-growing career. You can either start over with a purchased variety so that you know exactly what you have, or you can go the Zen route, accept that you now have no idea which garlic variety is which, and learn to live with some ambiguity in life.

When you're growing for market, however, ambiguity isn't exactly what folks are looking for.

"Hi, I'd like to buy some garlic! Can you tell me about this variety?"

"Actually, no. I have no idea which garlic this is because I f-ed up my harvest, but it's really well-grown and it tastes delicious. It's got a real garlic flavor to it. And it has cloves. Garlicky cloves, in fact. It's pretty awesome."

When growing a significant quantity of garlic, plan ahead. Know where and how you're going to hang it, and how you will keep varieties separate while hanging them. Know how you will plan to harvest so that only one variety is pulled at a time, reducing the chances of a mix-up. Make a plan for pulling, packing, and transporting garlic to its hanging spot in a manner that moves it quickly without getting bruised, crushed, or sunburned.

Throughout the rest of the harvest section, we'll share our tips and tricks. But every garden is different, and every small farm varies in infrastructure, and some of this shit you're just going to have to figure out on your own.

Digging

Hopefully, by now, we've established that garlic grows underground. Therefore, harvesting requires digging it up.

The key word here is *digging*. Note, please, that *digging* is not the same as *grabbing by the stalk, yanking as hard as you can and hoping for the best*.

Before pulling the garlic out of the ground, you need to snap off the roots anchoring it (rooting it, if you will) to the ground below. If you simply grab the stalks of garlic and pull, you're going to 1) work much harder than necessary, and 2) rip a lot of stalks from their heads.

The easiest way to snap the roots and loosen the garlic is with a potato fork.

Lies.

The easiest way to snap the roots and loosen the garlic is with a tractor and an under-cutter bar. But a tractor is overkill for most home gardeners, and it would take approximately 8,456 years to pay off a tractor if purchasing it with the profits from a small garlic farm. So it's probably best to resign yourself to the potato fork.[27]

When digging garlic, take great care not to damage it. Stick the fork in the ground two to three inches from the garlic stalk, and make sure it sinks deep enough to reach below the garlic head. Use the fork to push up the soil underneath the plant and the roots will snap, making it easy to grab the garlic by the stalk and pull it from the bed. Shake off the excess soil, but don't rub the heads to get every last bit of dirt off, as you will risk damaging the protective papers.

Stacking/Packing/Moving

Freshly pulled garlic should be treated with some care.

If you're harvesting just a few heads from your garden, it's fairly easy to pull them out, tie them up, and kick back with a beer to watch them dry (which will take anywhere from three weeks to a month or more, so monitor beer intake accordingly).

If you're harvesting larger amounts, however, two notes of caution:

1) Sunburn

Fresh garlic should be moved from direct sunlight as quickly as possible after harvest. Sunburn will affect both quality and storage life.

When harvesting garlic heads that numbered in the thousands, we would work down a row, pulling the plants until we filled a

---

[27] A potato fork is not a fork you use to eat a potato. DYHDS.

bin, and would then immediately move that bin to shade and begin again. Since we happened to use black plastic bins that heated up quickly,[28] we also tried to either keep the spare bins in the shade or keep their lids on them until we filled them so that we weren't putting the fresh garlic on a scalding hot surface.

We've said many times throughout this booklet that garlic is tough. It is.

But even the resilient among us don't mind a little extra love and care at times.

Especially if, just a few months down the line, it's our destiny to be chopped up and eaten.

2) Bruising

Another reason to be gentle with your young garlic harvest is to avoid bruising. If packing garlic in bins, it shouldn't be packed too densely or forced to fit.

Often, when we had friends come and help harvest, rather than drag the heavy bins with them as they worked, they would pull garlic heads and toss them in the general direction of the harvest bin with the plan to collect and stack them later.

As this had garlic heads both bouncing down the rows and scattered in the sun around the bin, such behavior produced some of my finer moments in garlic farming as I tried to gently correct our wonderful volunteers without screaming, shouting, or bursting into tears. I'm proud to say that not once did I lose my temper, nor did I throw myself down in between garlic rows and pound the earth with my fists. I also never ran around knocking over the harvest bins while shouting, "You wanna throw garlic all over the place? I'll show you throwing garlic all over the place!" I did not whack anyone in

---

[28] This is not an endorsement for black plastic bins as harvest vessels. It's what was available and affordable at the garlic-farmer price point. Dirt poor.

the head with a handful of garlic, nor, as a point of great pride, did I ever threaten a single soul with a potato fork.

I did think about all of these things, however. And I cried.

A lot.

But I was also hot, hangry, and exhausted for much of harvest season.

Be gentle with your garlic.

And your local garlic farmer.

Hanging Garlic

Garlic should be hung in a warm, dry location out of direct sunlight and where there is plenty of airflow. Locations like barns and sheds are an excellent choice. Garages often work well, too. For years, we simply hung our garlic under a massive second-story deck. Garlic will typically hang for three to five weeks, depending on your climate.

Again, for backyard gardeners handling a small harvest, hanging is pretty straightforward. Tie your garlic heads, greens still attached, in bunches of eight to ten, and hang them until the greens are dried through and/or the outside papers of the garlic are dry to the touch and flake off easily.

Drying large quantities of garlic takes more planning and additional infrastructure. Space is the first consideration, managing humidity and drying rates the second.

Condensing a substantial garlic harvest into a drying space usually means either planning for hanging multiple long chains of garlic, or using custom-built drying racks. *Custom-built* is, of course, shorthand for "whatever you can rig up yourself with your own personal level of carpentry skills."

Drying racks, while effective and attractive, take up a large amount of space. They also must be assembled and disassembled every year, and stored during the eleven months of the year when garlic is not being harvested. For us, racks seemed like an awful lot of work when we could just hang our garlic instead. We did rig up a couple of makeshift flat racks to dry our braiding garlic, but for the bulk of the harvest, we simply hung daisy chains of garlic from rafters using nothing but decking screws and baling twine. If you're game for redneck-winging-it like us, we've outlined that process for you here.

An overly detailed, step-by-step guide for garlic newbies so that you can't eff it up (even though you will anyway, because we all do):

1. Locate some solid rafters for hanging. These might be in a garage, shed, barn, or part of your neighbor's pergola.[29]

2. Screw in a series of screws along each rafter, approximately eighteen inches apart, leaving an inch or so of each screw exposed. Humid climates may need a little more space, dry climates a little less.

3. Sort bins of harvested garlic by variety and physically separate them so that you do not accidentally mix up varieties as you're stringing them.

4. Use the vast wisdom of the interwebs to review several videos on how to hang garlic using a daisy chain. This is not something easily explained in written word and it would be a waste of both of our time for me to try.

5. Hang one daisy chain of garlic. Each bunch of garlic should have around eight to ten heads, although this amount may

---

[29] Just kidding. Pergolas don't provide enough protection from the sun and not all neighbors appreciate trespassing.

vary due to climate.[30] Pro tip: bunching in groups of ten makes accurate harvest counts easy.

6. Label the daisy chain with variety name.
7. Repeat steps five and six until all the heads of garlic of one variety have been strung.

A word on labeling:

Some of you may just want to grow garlic to eat, and do not give a flying fig about what variety it is as long as it goes in your belly.

If you're an aspiring pro-grower, though, or just an anal-retentive OCD-er, we highly recommend using at least two types of labeling in case something goes wrong—e.g., the tape falls off the daisy chains, you lose your cheat sheet where you sketched out your hanging varieties, or the ink you used to label fades over the summer. These things happen.

We used three labeling strategies. Before harvest, we purchased colored duct tape, one color for every variety we grew that year. When we strung up the garlic, we assigned each variety a color. At the end of each daisy chain of garlic we placed a duct-tape tag, then doubled up by also writing the name of the garlic variety on the tag in permanent marker. Then we recorded a key of each color, plus its corresponding garlic variety, in our annual garden notebook.

If the notebook got lost, we still had the name of each variety written on the tags so we knew what the colors meant. If the print on the tags faded, we still had the notebook denoting the colors' varieties. If the tape fell off a string, we had a sketched map of the number of strings of each variety.

---

[30] Humid = fewer heads. Dry = more heads. If you haven't yet picked up on this theme, we need to work on your reading comprehension skills.

If, however, some asshole came over, cut off all the tape, rearranged all the strings, and burned our garden notebook while dancing naked around a bonfire, we were totally screwed.

That never happened.

Flat Racks for Braiding Garlic

Drying racks of every design can easily be found on the internet. There are racks built to hold hanging strings of garlic, much in the same way we hang from rafters. There are racks for drying garlic flat, racks for drying it upside down, and even, bizarrely, racks for sticking garlic out sideways to dry it at an angle, though that seems like a little overkill on the showing-off front.

As previously mentioned, we generally avoid racks altogether, with the one exception of setting up enough flat racks to hold our braiding garlic.

Allowing braiding garlic to dry flat makes it straighter and easier to work with when we braid it. The braiding process is easier because we're not wrestling with crooked necks, and the finished product looks cleaner.

For our braiding garlic, we have devised a highly technical drying system consisting of scavenged metal racks set on cinderblocks. We load each rack with enough garlic to have overlapping greens of up to two layers, but no overlapping heads. We set out a couple of oscillating fans to move air, then crack a beer and kick up our feet.

The garlic dries beautifully.

Drying Rules of Head (Garlic Doesn't Have Thumbs)

On average, drying your garlic down to the ideal storage level will take three to five weeks and possibly even longer in humid climates.

Knowing when to cut down and store garlic is almost as much art as science, unless you're a dorky scientist with multiple moisture-measuring tools, in which case it's probably all science and no art. Boring.

The garlic leaves need to dry until they are crunchy. The garlic heads should be dry enough that the outermost paper covering the head feels completely dry and papery to the touch. The roots, too, should be dry and brittle. When you unbind a group of the garlic hanging from a chain, there should be no green and no moisture left on the outer layer stems, even where they were tied together. The stems may still be somewhat pliable and not break when bent, but they should give all appearances otherwise of being completely dry.

If growing garlic in a dry climate like ours, be careful not to over-dry. Our first few years of growing, we found that our garlic was not storing as long as we'd hoped. It consistently fell short of the average storage lives for the cultivar types, so we took a close look at our growing practices and realized we were probably letting it hang too long. In an attempt to only store clean, market-ready garlic, we were leaving the garlic hanging until we cleaned it—a painstakingly slow process for a small farm of two. This meant that the last varieties we pulled down had easily been hanging six weeks or more in the hot, dry Colorado summers.

Once we realized what was happening, we switched up our technique. Rather than our clean-and-store process, we carefully monitored the hanging garlic. When it seemed sufficiently dry, we pulled it down, cut off the leaves and roots, and stored it without cleaning, choosing instead to clean it as we needed to for market. Not only did our garlic store much better that year, we also didn't waste time cleaning garlic that never went to market. Excess heads, or heads too small or too ugly to hit the farmers market table, were simply turned directly into garlic powder.

On the flip side, garlic growers in humid areas sometimes struggle to sufficiently dry garlic. Fans and dehumidifiers can help with garlic that refuses to dry. In the most humid environments or during extremely humid years, some growers move their garlic to climate-controlled garages, shops, or sheds where it's easier to control the ambient moisture. Even then, fans and dehumidifiers are often required.

A Note on Drying Braiding Garlic

If you are planning on braiding your softneck garlic, do not let it dry down completely before braiding. Once the stems[31] become dry and brittle, braiding is difficult and the workarounds suggested by some (such as wrapping the stems in a hot towel before braiding) are time-consuming and cumbersome.

Allow braiding garlic to dry down for about half the time you leave your non-braiding garlic. The leaves should be turning brown and you should see a notable reduction in the thickness of the stem. The outermost layer of the stem may be brittle and crack when bent, but the inner layers should still be pliable, allowing the stem to bend easily without breaking.

We found that the sweet spot for us to make braids was ten to fourteen days after harvest. If we began braids before ten days, the water content in the necks was still high enough that they shrunk noticeably after the braids were made, making for loose and sloppy-looking work. If we waited much beyond fourteen days, the stems became too brittle to easily work with.

---

[31] Technically, what I'm calling the stems of garlic aren't, scientifically speaking, stems. They're actually part of the leaves. But that's too damn picky to detail here. If it looks like a stem, braids like a stem, let's just call it a stem for the sake of a 101 level booklet, alright?

Cleaning and Storage

After ten months of prepping, planting, weeding, growing, harvesting, and hanging, you're almost done. You just have the last key steps: cleaning and storing the harvest. It's about time, too, because it's probably already August and you're going to start this whole party over again in October. So, if you want to catch any sort of break at all, you better bust through the rest of this mess fast.

If you've gotten this far only to realize that you're over gardening, that it takes entirely too much work, and that you just want to get your summers back so you can go to Mexico, sip on a margarita, and forget that you ever dabbled in garlic growing, cleaning should be pretty easy. Cut off the garlic stems, chuck your garlic in a basket, and store out of direct sunlight and away from heat sources. Purchase a ticket to Mexico and order a margarita.

If, on the other hand, you've gotten this far and feel you've invested entirely too much work to cut corners on the very last steps of a nearly year-long process, and you're determined to see your garlic through until it positively glows with perfection in your Martha-Stewart-inspired kitchen, we have a few last tips.

First off, in solidarity with the Fuck-It-I'm-Going-to-Mexico crowd, truly, don't clean if you don't have to. As we discovered when we stopped over-drying our garlic, cleaning every head before storage is unnecessary and potentially detrimental. The papers around garlic are what protect the cloves. Keeping them intact allows for longer shelf life and saves you boatloads of time.

Cut the stems and pack in baskets, milk crates, or vegetable crates—any container designed for plenty of airflow. Store, uncovered, in a cool (not cold) location. In terms of storage temperatures, as with so many things, garlic is quite forgiving. The internet weirdos will try and convince you otherwise. In a five-minute Google search to check my personal experience against the collective interweb wisdom, I found impossibly contradictory claims.

"Commercial garlic storage aims to keep bulbs between 56 and 58F. Less than 50 F, your garlic is likely to sprout. Don't be tempted to store garlic in your fridge! Above 66 F, your garlic will quickly shrivel."—Cornell College of Agriculture and Life Science.[32]

"For long-term storage, garlic is best maintained at temperatures of 30 to 32 °F with low RH [relative humidity] (60 to 70%)." — University of Massachusetts Amherst's Center for Agriculture, Food, and the Environment.[33]

"Softneck garlic can be kept in good condition under commercial storage conditions for up to 9 months when held near 32°F (0°C), or for 1 to 2 months at ambient temperatures of 68° to 86°F (20° to 30°C). Hardneck garlic will store for up to 6 months under ideal conditions."—University of California, Agriculture and Natural Resources.[34]

Then, of course, other garlic growers recommend 35–55°F, 60–65°F, or 50–60°F, and seriously, internet, why can we still not agree on what should be a testable scientific fact?

Perhaps it's because garlic is so agreeably resilient.

For the entire duration of our farmers market years, we stored our market garlic in the basement in baskets where temps usually ranged between 65 and 70°F. It was fine. The softneck garlic braid that has been hanging by our kitchen cabinets for the last

---

[32] Petra Page-Mann, "Secrets of Storing Garlic," Cornell Small Farms, July 20, 2020, https://smallfarms.cornell.edu/2020/07/secrets-of-storing-garlic/.

[33] Ruth Hazzard, "Garlic Harvest, Curing, and Storage," University of Massachusetts Amherst, last modified January 2013, https://ag.umass.edu/vegetable/fact-sheets/garlic-harvest-curing-storage.

[34] Linda J. Harris, "Garlic: Safe Methods to Store, Preserve, and Enjoy," University of California Agriculture and Natural Resources, October 2016, https://doi.org/10.3733/ucanr.8568.

ten months is also still good. Yes, temperature will impact your garlic's storage life, but so will things like how you dried it, how you cleaned it, the number of papers left on the bulbs, each variety's unique characteristics, and whether or not your garlic likes that smug look on your face.

The one thing that nearly everyone agrees on is that you want to keep garlic out of refrigerator-like temperatures, which typically run 33–44°F. Remember that garlic is a fall-planted crop that uses cold as a signal to sprout and get established before the ground freezes for winter. Keeping garlic at refrigerated temperatures encourages sprouting. Likewise, keeping it in a hot environment degrades it.

Keep it cool, then cool your jets.

Your garlic will do just fine.

When you are ready to clean it, be it for selling at a farmers market, or for keeping a small basket of ready-to-go goodness on your kitchen counter, gently rub off the crinkly, loose outer layer of papers until you get to papers that are still tightly attached to the cloves. As you do this, take care not to remove unnecessary layers. Have some papers that come off partially but then stick? Grab an old toothbrush, and with the bristles at an angle, gently brush against the stuck paper. The bristles will snag it and easily remove only the desired layer.

Rub off any excess dirt remaining on the roots, then trim the roots with a pair of scissors. Stack in a basket or pack in a bin. Congratulations.

You are a bonafide garlic farmer.

CHAPTER 6

FUCK YOU, FUSARIUM

Weeding my way through the garlic rows in late spring one year, I reached the last row in our plot and stepped straight into a horror movie.

The plants were crawling with grasshoppers. They were positively covered in them, the hoppers happily munching away, demolishing our future harvest with their nasty little jaws.

Grasshoppers are insects from the Acrididae family and, evolutionarily speaking, direct descendants of Lucifer himself. Some of the most destructive and spiteful insects on God's green earth, grasshoppers have stumped even the professional advice-givers on university extension websites. Google *grasshopper control* and your computer will give a low whistle, shake its head, and mutter, "Whoooo-eeee. You got yerself a 'hopper problem? Welp, my friend. You are screwed."

Throughout the years we grew for market, I engaged in a battle of wits against these sadistic assholes. They ate my rhubarb and decimated the beans. They turned heads of cabbage into lace and annihilated our hops. They ate the tops off the onions, the petals off the sunflowers, and then, for an encore performance, they shat all over the kale.

As insatiable as they were, however, they had never, not once, touched the garlic.

Yet here I was, staring at a garlic row covered in the bastards merrily downing a plant they generally don't like. Something wasn't right.

I brushed off the grasshoppers, grabbed one of the plants, and pulled. The garlic head came up, but the basal plate and the roots stayed in the ground.

Uh-oh.

I pulled another plant. It came up whole, but the bottom was red and pockmarked, obviously inedible. I pulled several more plants. They all showed signs of the same demise: basal rot, also known as Fusarium, had infected that entire variety.

The grasshoppers were the symptom of a problem, not the cause. Pests gravitate toward plants whose defenses are already weak. If, in seven years, I'd never lost a garlic plant to grasshoppers, seeing an entire row suddenly coated in them was about as subtle a sign that something was wrong as, say, waking up one morning to discover you have no legs. Presuming, naturally, that you went to bed with legs. You wake up in the morning, and just have this feeling... *Wait a minute, something's off...* then you roll out of bed and *blammo!* hit the floor. *Ah, well, there it is. It would seem I have no legs.*

Pulling our Fusarium-infected garlic provoked a similarly resigned revelation. Just to be sure, I selected a few of the hardest-hit heads and had Mike send them to the Colorado State University

Plant Diagnostic Clinic. When he called some weeks later to get the results, the plant diagnosticians were positively giddy.

"Oh, you sent the garlic in! Yes! Yes, it was Fusarium. In fact, during our staff meeting every month, one of the topics we discuss is, *What's the worst specimen to come in this month?* and this month it was your garlic! A really terrible case of Fusarium! It looked awful! Glad we could help!"

They were so damn cheery that I'm pretty sure someone in the lab had a beer bet riding on our garlic. *Pssht, you think that barley plant you got looks rough? I'll bet you a beer I found something worse.*

Our one Fusarium year aside, however, we have been remarkably disease free in all other garlic-growing years. Just as it is resilient and hardy in most growing conditions, garlic is also fairly disease resistant as long as you include a proper growing rotation and basic care. Still, it is worth mentioning briefly a handful of the most common issues that can arise when growing garlic.

Common Garlic Diseases and Pests

Fusarium Basal Rot

The year of the garlic-eating grasshoppers was a direct result of Fusarium basal rot. A fungus that lives in the ground and occurs worldwide, it can invade any allium crop at any stage of growth when conditions are right. Symptoms of Fusarium basal rot usually include yellowing or leaf dieback, beginning at the tips of leaves and working downward. The actual rot will begin at and around the basal plate, and is recognizable by its reddish-brown color. The rot will progress toward the stem, and infected cloves will be brown and watery.

Fusarium can become visible after harvest while garlic is in storage. If you find watery brown cloves in a stored batch of garlic, Fusarium may be your culprit.

As Fusarium is present everywhere and can survive indefinitely in soil, the best protection is to provide ideal growing conditions for garlic. Happy and healthy garlic resists disease. Fusarium enters the plant via damage to bulbs, so take care in planting not to nick or cut cloves, and do not plant damaged cloves. Keep a minimum of a three- to four-year rotation on plots planted with alliums.

You can also plant disease-resistant varieties. While we did not specifically seek out Fusarium-resistant varieties, it became quickly evident which varieties were susceptible to it. The Year of the 'Hoppers, Fusarium absolutely obliterated one variety—and only that variety.[35] We chucked every last bulb of it in the trash and never grew it again. Only one other year did we pull another variety that showed signs of Fusarium. Again, only that single variety showed symptoms, and while we were able to save some of that batch for culinary purposes, we did not replant any of our seed stock.[36] The rest of our garlic varieties performed—and continue to perform—well despite the significant presence of Fusarium that we know exists in our soil.

More Funguses! Or Fungi? Fungees?

More nasty shit that comes from spores!

While Fusarium warranted a section of its own due to its worldwide presence, interminable lifespan, and the ability to develop in garlic at any stage of life, fungal diseases as a whole are the number one issue when it comes to garlic pathogens. Which fungal disease you have, however, depends on any variety of factors including your growing region, the weather, your soil, and shit luck.

---

[35] You're going to ask, so I'll just tell you. It was the Bogatyr variety.
[36] Spanish Roja. Sexy sounding. No bueno performance in our soil.

I am not a garlic disease pathologist.

I am, however, a big believer in doing your homework (I believed we talked about that already). When you find an issue in your garlic patch, a quick read of a garlic-growing booklet isn't going to be enough. Identify the most notable characteristics of what you see going on, and start doing some (metaphorical) digging. Send a sample to a university extension office as we did. They'll test it, and the cost is cheap. As a rule, fungal diseases are a good jumping-off point due to their prevalence in garlic. Here's a short list of some of the most common ones:

DOWNY MILDEW:

Pale splotches on leaves with a fuzzy whiteish-purple, mold-looking growth.

RUST:

White spots on leaves that develop into orange pustules.

BOTRYTIS NECK ROT:

Appears late in the garlic life cycle (late spring, early summer) and is initially identified by rot of the neck of the plant at the soil line. Fairly common.

WHITE ROT:

The Grim Reaper of alliums. Stunted plants, yellowing and wilting of leaves, and progressive degeneration and rotting through the rest of the plant. There are no known controls or treatments. If you get white rot, you are completely fucked and cannot grow any allium plant in that space again for at least twenty years, or maybe for the rest of your life. You should sterilize the shit out of any tool you ever used in an infected area, then burn everything to the ground and move away, forever.

Other Common Issues

GARLIC MOSAIC VIRUS:

A catch-all name for a group of viruses in the genus Potyvirus, garlic mosaic virus is recognized by a mottling or striping of light green colors on the leaves. Symptoms are usually mild but can include stunted plants or reduced bulb size. There is no treatment once a plant is infected with garlic mosaic virus. As transmission is through seed stock, infected plants can be pulled out and destroyed or, if symptoms are mild, separated at harvest so they are not used as stock.

GARLIC BULB MITES:

Garlic bulb mites can cause damage both in growing garlic and stored garlic. Capable of overwintering in soil, garlic mites will nibble on the bulbs underground, but symptoms above ground can be hard to catch or easily mistaken for other garlic diseases. Mite infestation causes yellowing of leaves or brown tips on the garlic, both of which are common. Underground, however, mites frequently live and throw huge, raging mite parties where the basal plate meets the bulb. Bulb mites can cause stunted plants and weak root systems, but just as problematically, their bite marks can be an entry point for Fusarium and other diseases. They can also survive in stored garlic, continuing their little mite frat parties and damaging your stored stash until it's nothing but stanky, rotting mush.

Disease Prevention

While by no means a comprehensive list, the above are some of the more common issues that can be found with garlic. Almost all of these diseases can be avoided by three straightforward actions: 1) Purchase clean seed stock, tested for common garlic pathogens. 2) Rotate your garlic plot on a minimum of a three-year rotation, ideally four years. 3) Give your garlic good growing conditions: fertile soil, appropriate watering, and regular weeding. Don't stress it out.

None of us grow well when stressed.

A little love can go a long way.

CHAPTER 7

USE THAT SHIT!

One of garlic's greatest strengths is its storage capacity. Fresh heads last for months, but even they can be dried down into garlic powder, making their storage life indefinite.

When we began growing garlic in large quantities, I quickly realized that we would need a plan for all the garlic heads that were not farmers market quality. Even in the best harvest year, a certain percentage of garlic is going to be too small or misshapen to put on display, and even with great care, some heads are always damaged during harvest.

Unable to bear the thought of garlic going to waste, I spent many an hour in our kitchen those first few summers, figuring out the best way to process all of that "reject" garlic into something both storable and tasty. Processing garlic from a whole, freshly harvested head into a perfectly sifted powder, however, is a multi-step process that, as I later realized, is best executed with a little forethought and planning. As opposed to haphazard winging and flinging.

Unfortunately, I'm more of the wing-and-fling type. As a result, my garlic processing adventures, while highly educational and informative, occasionally deviated from what one might consider a well-thought-out process.

FACT:

Dehydrating garlic indoors will stink up the house so badly that you might find yourself debating whether the best manner of eliminating the smell would be to a) call a cigarette smoke mitigation company and see if they could do anything about garlic, or b) burn the house down.

FACT:

If you handle too much raw garlic without the appropriate caution, you may develop an allergic skin reaction to garlic oil.

FACT:

Processed garlic dust can make you sneeze for several minutes straight, at times to the point of hyperventilation.

FACT:

Garlic papers, if not removed from any and all surfaces while still moist, will adhere to said surfaces like Gorilla Glue and remain there for-fucking-ever until you have to sell the house and seriously consider just painting over glued-down garlic papers and calling it texture.

FACT:

Simply pulling out a dehydrator and a pile of raw garlic bulbs may cause a spousal domestic exodus, possibly accompanied by enthusiastic profanity and some variation of the phrase, *Not this stinky-ass bullshit again.*

As one might anticipate with such a highly refined research and development process, and such a clear track record of attention to detail, my garlic powder experiments were a huge success.

They were also a huge mess, but over the years, I refined my techniques, reduced the chaos, and developed several hacks to make the process easier. I also started blending our garlic powder into garlic salt concoctions made with other herbs and ingredients from our garden, anticipating that different garlic offerings would appeal to different customers and help us boost overall sales.

To my surprise, sales of our garlic powder and garlic salts quickly outpaced those of fresh garlic. I had no idea that people love seasonings so much. We had die-hard fans who came to the market each week to purchase our fresh garlic, but in a cage match, the fresh garlic fans would never stand up to the garlic salt fans. Folks who purchased jars of our salts while on vacation in Colorado started contacting me from out of state, wanting to know how they could order more. Regular market stand customers would place their Christmas gift orders in July so they could be assured we wouldn't run out. Friends and family would pay me for salts before we even began garlic harvest.

People are willing to put up with a lot of bullshit in this life, but bland food? Never.

When we realized that we no longer had the capacity to continue producing at a commercial level, one of the hardest things I had to do was tell people that this would be the last batch of garlic salts they would be able to purchase. And so, when writing this pamphlet, after a brief debate with myself about competition, intellectual property, and other proprietary bullshit, I made the easy decision to share my entire process here, along with all of our most successful garlic salt recipes.

Use them all. Make the recipes yours. Share them with others. Help us keep some vestige of our little farm moving forward throughout the garlic generations.

Or just sprinkle some on your pizza and enjoy.

Garlic Powder Processing Equipment

The basics required to turn garlic into garlic powder include:

- Food processor
- Food dehydrator
- Vitamix OR coffee grinder
- Mason jars or Tupperware for storage
- Fine mesh sifters
- Latex or nitrile gloves
- Respirator or N95-type masks
- All your standard kitchen utensils used for processing any produce: cutting boards, knives, spoons, bowls, etc.

A few notes on a few of the above items:

Food Dehydrator:

Make sure you have a food dehydrator that has a fan for blowing hot air as well as temperature controls. The cheapest dehydrators have only heating elements without fans. If you attempt to use this type of dehydrator for garlic, you will be waiting many days for a single batch of garlic to dry.

Additionally, different brands of food dehydrators come with different types of trays. If possible, avoid purchasing a dehydrator with inflexible plastic trays. Garlic (and many other dehydratable items, such as tomatoes) sticks to those trays and is difficult to remove.

Instead, look for a dehydrator that has bendable mesh trays, which make popping off the dried garlic quite easy.

If you already own a dehydrator with inflexible trays, putting a layer of parchment paper down before spreading the garlic can make the process a little easier, though sometimes the garlic will stick to even the parchment paper.

Vitamix:

For processing large amounts of garlic powder, a Vitamix is a game changer and, although pricey, is well worth the money. For home gardeners, however, if you don't have a Vitamix, a coffee grinder will do the job, just in much (much) smaller batches.

Fine Mesh Strainer or Sifter:

At a minimum, you want a strainer (or sifter) with a mesh fine enough that you could use it for sifting flour. Again, if you are processing relatively large quantities of garlic, another worthwhile time investment might be a set of stackable sifters between ten to fourteen inches in diameter. These will allow you to easily sift large quantities of garlic much faster.

Latex or Nitrile Gloves:

Garlic oil is surprisingly potent stuff. It's one thing to peel a few cloves of garlic for dinner. It's another thing to be dealing with dozens (or hundreds or thousands) of heads that have been processed to release oils. After my first batch of processing garlic without gloves, I got a wicked case of eczema on my hands and quickly realized that for handling large quantities of garlic, gloves were the way to go. This is also true for handling garlic powder.

Respirator or N95 Masks:

When you make garlic powder you will sneeze. Incessantly. Should you choose to make jalapeño garlic powder as per the recipe provided, you will sneeze, snort, cough, choke, cry, and probably insult me for even having suggested the idea of such a recipe. Years before COVID made them cool, I was using masks to get through garlic processing. While medical-grade masks do well with garlic powder, the only way I could ever get through the jalapeño garlic salt processing was with a respirator.

Processing Garlic Powder

1. Prep Garlic Heads

The most key, sanity-saving rule of making garlic powder is to not waste your time peeling the individual cloves.

Every part of the garlic plant is edible, from the leaves to roots. While you may not want to put all of these parts into your garlic powder, on a practical level it means you don't have to peel your cloves, which will save you approximately 8,584 hours of exasperatingly sticky and tedious labor.

To prep the heads for processing, brush off all loose layers of garlic papers. Be sure to remove any layers that have residual dirt on them to avoid an unpleasant grit in your powder. Then, rub off any papers that are loose and easily removed. Although these are edible, not having loose papers will make the garlic easier to work with.

If you are processing softneck garlic, simply cut off the basal plate using a sharp knife, then cut the head into halves or quarters. This makes it process more smoothly in the food processor.

If you are processing hardneck garlic, it's a little more work due to the hard stem[37] running through the center of the heads.

Airplane black boxes should be made from those stems. They are indestructible. Food processors won't touch them: you'll just pull out a mush of garlic punctuated with completely intact, unscathed hardneck stems. Even the Vitamix, which runs on speeds that pulverize gravel, struggles with these hardneck stems. It sort of grinds them into large shards that, if they end up in your garlic powder, have a distinctly stick-like firmness.

If you have an extremely sharp kitchen knife, you can cut off the basal plate of hardneck garlic, just like the softneck, but this requires cutting through that stem, which is both difficult and somewhat dangerous, as it's easy for the knife to get stuck in it and cause your hand to slip.

Over time, my preferred method of dealing with hardneck heads has been to simply stick the blade of a sharp knife between two cloves of the head, pop them apart, then use my hands to break off the rest of the cloves into a bowl.

2. Make Garlic Mush

Fresh garlic heads must be processed into a garlic pulp *before* being dehydrated. Dehydrating whole cloves creates small, rock-hard pieces that even a Vitamix will struggle to grind.

To make this mush, take the cleaned garlic and run it through the food processor at high speed until all individual cloves are chopped into bits. The consistency of the mush is not super important. The key is to ensure that all the garlic cloves have been broken down and no whole ones are left in the mix.

---

[37] Not the scientifically correct term. If it looks like a stem and acts like stem, call it a stem.

3. Spread Mush Over the Dehydrator Trays

If you have flexible mesh trays, spread your garlic mush straight over the trays. It will be thick enough to not fall through the holes.

If you have rigid plastic trays, I recommend lining them with parchment paper first, then spreading the garlic over that, as it will make the removal process easier after dehydration.

Keep the garlic as evenly spread over the trays as possible, as this will help with the uniformity of the drying process. A good rule of thumb is to spread the garlic about one quarter inch in thickness.

4. Dehydrate at 100–125°f For… as Long as It Takes

Heat breaks down the enzyme alliinase, a key component in creating allicin, which, you may remember, is one of the most beneficial therapeutic compounds found in garlic. To retain as much alliinase content as possible in our powders, we always dehydrate at 125 degrees Fahrenheit or under. However, the lower you go temperaturewise, the longer it takes to dry.

Overall, the time required to dry your garlic will depend on a variety of factors: your dehydrator, the ambient humidity, how thickly you layered the garlic on the trays, how fresh or aged the actual garlic heads are, and others. Garlic that was just harvested will take the longest to dry, as the moisture content is still extremely high. Garlic that has already been cured and stored will have lost some of that moisture content and dry much faster.

As a starting point for all of our full dehydrator batches of garlic, I start with sixteen hours. You want the finished garlic to be uniformly dry with absolutely *no* moist spots. Moisture causes spoilage. When you pull the garlic off the trays it should snap, not bend, and even the most thickly layered sections should be dried completely through.

A word to the wise: do not dehydrate garlic inside your house. The smell will be overwhelming and the entire house, curtains to carpet, will absorb that garlicky smell and leave it lingering for weeks to months. Likewise, I even recommend against dehydrating in an attached garage, as the smell will still make it into the house, albeit much less potently.

If you're making garlic powder for home use, the best option is to dry it outdoors on a covered patio or in a small shed—anywhere protected from the elements.

If you're making powder to sell commercially or via a cottage food law, most state food regulations won't allow outdoor food production, in which case you're screwed and whatever kitchen you use—plus everything in about a ten-block radius around it—will smell like garlic.

Bring lots of bribe money for anyone sharing the space with you.

5. Grind and Sift the Powder

Once the garlic is dry, pull the trays out of the dehydrator and peel the garlic off of them. It will come off in large, rough chunks. Break these bits up into smaller chunks and fill a Vitamix.

Run the Vitamix on high until you have a fine powder. Sift the powder that comes out of the Vitamix using a fine mesh sifter in order to create a uniform final product. Garlic pieces too large to make it through the sifter can be returned to the Vitamix for a second grinding.

If you are using a coffee grinder rather than a Vitamix, the instructions are the same, just in much smaller quantities.

6. Store

Keep the powder in an airtight container out of direct sunlight and away from heat sources.

We often got the question at market, "How long will your garlic powder last?"

Think of your spice cabinet at home with its jars of spices, some of which I would bet an entire year's garlic harvest has been sitting in for years on end. Spices and spice blends don't go bad, per se. They simply lose their potency over time, becoming less notable in flavor.

Garlic Salt Blends

The following are recipes for the five most popular garlic salt blends that we sold.

Because we were a small farm and grew a variety of vegetables and herbs in addition to garlic, all of the ingredients listed in these salts were also grown on our small farm. If you choose to do the same, a few quick notes on dehydrating and processing the additional ingredients:

Garlic Scapes:

The process for dehydrating and processing the scapes is exactly the same as for garlic heads. The end result will be a light green powder with a milder taste than that of the powder made with garlic cloves.

Jalapeños:

Halve the jalapeños and set them on the dehydrator tray, sliced side up. If you want to tone down the heat factor but keep the jalapeño flavor, cut out ribs and seeds. Dry at roughly 125 degrees Fahrenheit until the peppers snap when broken in half. Process in Vitamix or coffee grinder. **Process outside and wear appropriate protective gear.**

Herbs:

Dry herbs at lower temperatures of around 95 degrees Fahrenheit until brittle. Strip dried leaves off stems, and process in Vitamix or coffee grinder.

Spring Salt

Spring salt is made with the garlic scapes harvested from hardneck garlic in late spring. Our personal favorite of all of the garlic salts we produced, spring salt acts as an excellent flavor enhancer, giving food an extra depth of flavor without making it overwhelmingly garlicky. At home, we replaced our table salt with spring salt and used it on everything.

Ingredients:

  Sea salt
  Garlic scape powder

Instructions:

Blend two parts sea salt to one part garlic scape powder. Mix thoroughly. Store in an airtight container out of direct sunlight and away from heat sources.

Garlic Salt

Every garlic farm must have a standard garlic salt recipe. This was ours.

Ingredients:

  Sea salt

Garlic powder

Instructions:

Blend three parts sea salt to one part garlic powder. Mix thoroughly. Store in an airtight container out of direct sunlight and away from heat sources.

Jalapeño Garlic Salt

Great on pizza, popcorn, eggs, or anything else that needs a little heat.

Ingredients:

Sea salt

Jalapeño powder

Garlic powder

Instructions:

Blend three parts sea salt to one part garlic powder and one part jalapeño powder. Mix thoroughly. Store in an airtight container out of direct sunlight and away from heat sources.

Rosemary Garlic Salt

Excellent on fish, poultry, potatoes, and artisanal breads.

Ingredients:

Sea salt

Rosemary powder

Garlic powder

Instructions:

Blend three parts sea salt to one part garlic powder and one part rosemary powder. Mix thoroughly. Store in an airtight container out of direct sunlight and away from heat sources.

Roasting/Grilling Salt

Slightly trickier to make, Roasting/Grilling Salt is well worth the extra effort. A close second to Spring Salt, we put Roasting/Grilling Salt on...everything we roasted or grilled. Mixed with a couple of tablespoons of a good olive oil, this salt is excellent on grilled vegetables or oven fries. Use it to coat a whole chicken before roasting, or to marinate steaks.

This is intended to be a chunkier salt, which gives great bursts of flavor when used as a marinade but requires processing the garlic and herbs less than for the finer garlic salts. For the garlic in this, I used the chunks sifted out from the finer powder. These chunks should be roughly Kosher salt size to produce the best uniformity.

The herbs for this recipe are processed for shorter intervals in the Vitamix to produce a chunkier result that blends well with the Kosher salt.

Ingredients:

   Kosher salt

   Chunky garlic powder

   Roughly ground rosemary

   Roughly ground sage

Instructions:

Blend one part rosemary, two parts sage, and four parts each of Kosher salt and chunky garlic powder. Mix thoroughly. Store in an airtight container out of direct sunlight and away from heat sources.

CONCLUSION
CONGRATS ON FINDING YOUR PLACE IN THE UNIVERSE

One day during our garlic-growing adventure, with crates of garlic stacked in our basement and our back patio converted into a garlic cleaning facility,[38] I went to snag more heads for our market stash that weekend. Entering the basement, I noted a distinct funk.

I sniffed around the basement, looking for a dead mouse, a dog turd courtesy of our pooch, or some other clear and inevitably gross source of the smell. But I found nothing. Shrugging, I grabbed a crate of garlic and took it outside. Halfway through cleaning my allotted batch, I came across a head of garlic with one watery, brown clove that smelled funky.

---

[38] By which I mean it had a table with scissors, toothbrushes (for peeling off wrappers, duh, did you even read Chapter 5?), loads of baskets, and a hot friggin' mess of garlic wrappers.

It was a little bit of Fusarium in the stored garlic, a not-uncommon occurrence. One mushy clove out of an otherwise healthy batch of heads didn't worry me much. What surprised me, however, was the fact that I had sniffed out the stink.

Concerned that maybe the other crates were filled with Fusarium-rotted heads, I went to the basement and pulled them all out, digging through each and squeezing heads at random to see if any were soft or softening. My search produced nothing but healthy heads, so I wrote it off as a fluke and resumed cleaning.

Over the course of that season, however, and all the seasons following it, the same experience happened repeatedly, and I realized that I had developed an ability to sniff out a handful of bad cloves within entire crates of otherwise good garlic. My otherwise unimpressive schnoz continued to refine its freaky garlic sniffing abilities so that I could tell you the difference in smell between freshly harvested garlic and garlic that had been hanging for a while. I could tell when the garlic in the dehydrator had been left in there too long by a subtle acrid quality to its aroma. I could distinguish whether someone was working with fresh garlic or garlic powder, just by walking in the kitchen. I could tell when various varieties of stored garlic were starting to sprout because their smell changed.

At first, I thought I was some sort of garlic savant. A gifted garlic goddess. A sensitive, smelling seductress. Then, I talked with a farmer friend who raised pastured poultry.

"Ugh," she told me. "This is really gross, but a bunch of my chickens caught a virus from wild birds and while they're going to be all right, I've realized I can tell which birds are sick just by their smell. Sometimes I can tell even before they have symptoms."

Turns out that when you spend a lot of time with something, you become in tune with parts of that something in ways you never expected.

Grow garlic, learn to smell rot.

Or rather than being able to sniff out a fungal disease, perhaps you'll find you simply learn a little more about the land that you live on. In my years of growing garlic, I have had the pleasure of experiencing the pulse of life that is always happening around us but is so often out of sight and out of mind.

One year, a robin followed me around while I tilled the field. She grabbed worm after worm from the freshly turned soil until she had twenty worms hanging from her mouth. I had a serious case of the giggles just watching her pack her beak full.

Another year, while weeding, a little head popped up out of the earth right next to my hand. A mama rabbit had made her den underneath a row of our hardneck garlic, and four little furry babies stared at me. I hate rabbits in the garden, but the babies were so cute that they got a free pass for the season. The rabbits generally left the garlic alone, anyway.

I learned which trees in our yard had the Blue jay nests, and watched when the baby jays learned to fly. My heart jumped out of my chest, worried that their experiments in aviation would fail, and knowing that the way of Mother Nature is that some inevitably do.

I found a Monarch caterpillar on our tiny milkweed patch, coyote scat along our fence line, and a fox den under our neighbor's shed.

Our last year of growing for market, Mike and I realized that migrating bats were living under the eaves of our garage, a family of horned owls in the trees behind the house, and toads in our perennial flower patches. We sat on the back patio every evening, waiting for all of them to come out of hiding, gearing themselves up for their nocturnal adventures.

Growing garlic is just one specific task. You take a few cloves, plunk them in the ground, throw on some water and compost, and a few months later you're congratulating yourself as you chop up a few of your homegrown cloves to make a batch of pesto or a loaf of garlic bread.

But growing garlic—growing anything—is also a humbling experience. The plant's survival is dependent on your care, yet you're also at the mercy of the whims of weather, of nature. You labor nine months to produce these little vegetables that will fit in the palm of your hand, then you stop and wonder, just for a moment, that if this much time, work, and effort goes into growing a baseball-sized plant, then how much time, work, and effort should you be putting toward yourself as a full-sized human?

Yet, somehow, in the garden, these musings don't seem too heavy. With your hands in the soil, the breeze rustling through the plants, and the sun heavy on your back, you understand, with newfound clarity, that you are now amongst those who know, without a shadow of a doubt, that garlic does not grow on trees.

Bibliography

Ankri, Serge, and David Mirelman. "Antimicrobial Properties of Allicin from Garlic." *Microbes and Infection* 1, no. 2 (February 1999): 125-129. https://doi.org/10.1016/S1286-4579(99)80003-3.

de Menenzes Torres, Kátia Andrea, Sônia Maria Rolim Rosa Lima, Luce Maria Brandão Torres, Maria Thereza Gamberini, and Pedro Ismael da Silva Junior. "Garlic: An Alternative Treatment for Group B *Streptococcus*." *Microbiology Spectrum* 9, no. 3 (November 2021). https://doi.org/10.1128/Spectrum.00170-21.

Engeland, Ron L. *Growing Great Garlic: The Definitive Guide for Organic Gardeners and Small Farmers*. Okanogan, WA: Filaree Productions, 1994.

Gómez-Alonso, Juan. "Rabies: A Possible Explanation for the Vampire Legend." *Neurology* 51, no. 3 (September 1998). https://doi.org/10.1212/WNL.51.3.856.

Harris, Linda J. "Garlic: Safe Methods to Store, Preserve, and Enjoy." University of California Agriculture and Natural Resources. October 2016. https://doi.org/10.3733/ucanr.8568.

Hazzard, Ruth. "Garlic Harvest, Curing, and Storage." University of Massachusetts Amherst. Last modified January 2013. https://ag.umass.edu/vegetable/fact-sheets/garlic-harvest-curing-storage.

Li, Guoliang, Xudong Ma, Lisha Deng, Xixi Zhao, Yuejiao Wei, Zhongyang Gao, Jing Jia, Jiru Xu, and Chaofeng Sun. "Fresh Garlic Extract Enhances the Antimicrobial Activities of Antibiotics on Resistant Strains *in Vitro*." *Jundishapur Journal of Microbiology* 8, no. 5 (May 2015). https://doi.org/10.5812/jjm.14814.

Meredith, Ted Jordan. *The Complete Book of Garlic: A Guide for Gardeners, Growers, and Serious Cooks*. Portland, OR: Timber Press, 2008.

Page-Mann, Petra. "Secrets of Storing Garlic." Cornell Small Farms. July 20, 2020. https://smallfarms.cornell.edu/2020/07/secrets-of-storing-garlic/.

Palaksha, M. N., Mansoor Ahmed, and Sanjoy Das. "Antibacterial Activity of Garlic Extract on Streptomycin-Resistant *Staphylococcus Aureus* and *Escherichia Coli* Solely and in Synergism with Streptomycin." *Journal of Natural Science, Biology, and Medicine* 1, no.1 (July 2010). https://www.ncbi.nlm.nih.gov/pmc/articles/PMC3217283/.

Volk, Gayle M., and David Stern. "Phenotypic Characteristics of Ten Garlic Cultivars Grown at Different North American Locations." *HortScience* 44, no. 5 (August 2009). https://doi.org/10.21273/HORTSCI.44.5.1238.

www.ingramcontent.com/pod-product-compliance
Lightning Source LLC
Chambersburg PA
CBHW061338040426
42444CB00011B/2983